Cabin Crew Conference 2019

Identifying and Developing Professionalism in Aviation Training

Orlando, Florida, USA
30 April – 2 May 2019

ISBN: 978-1-7138-1889-2

Printed from e-media with permission by:

Curran Associates, Inc.
57 Morehouse Lane
Red Hook, NY 12571

CURRAN ASSOCIATES INC. proceedings.com

Some format issues inherent in the e-media version may also appear in this print version.

Printed with permission by Curran Associates, Inc. (2020)

For permission requests, please contact Halldale Media Inc.
at the address below.

Halldale Media Inc.
735 Primera Blvd.
Suite 220
Lake Mary, Florida 32746
USA

Phone: (407) 322-5605
Fax: (407) 322-5604

www.halldale.com

Additional copies of this publication are available from:

Curran Associates, Inc.
57 Morehouse Lane
Red Hook, NY 12571 USA
Phone: 845-758-0400
Fax: 845-758-2633
Email: curran@proceedings.com
Web: www.proceedings.com

TABLE OF CONTENTS

SOUTHPAC
aerospace

FROM REACTING TO LEARNING…THE NEXT STEP IN AVIATION SAFETY

1. Where are we now? Reacting?
2. Current Methods
3. A “New View” of Safety
4. Learning Methods

CONTEMPORARY SAFETY?

COMPLIANCE MONITORING

DASHBOARD SAFETY – CRAZY METRICS & RIGID RULES

You've heard it said,
"what gets measured, gets done?"

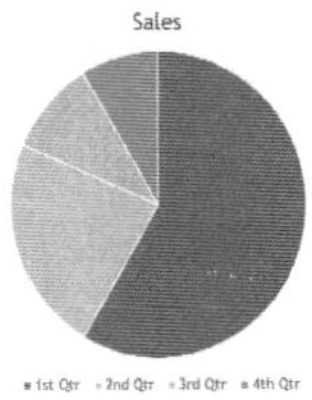

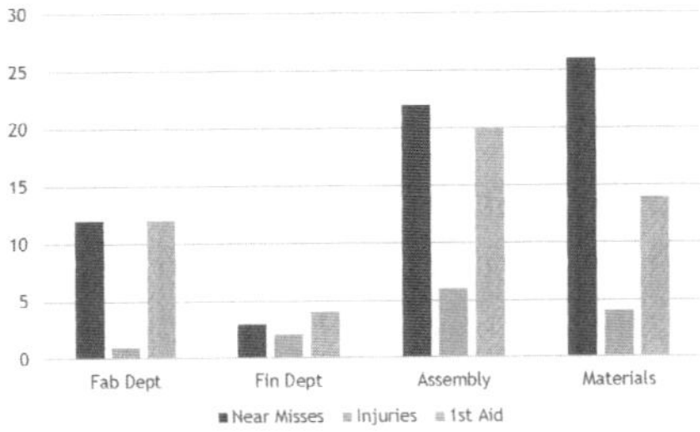

International Standards
and Recommended Practices

Annex 13
to the Convention on
International Civil Aviation

Aircraft Accident and
Incident Investigation

COMMON 'CAUSES' AND FIXES

CAUSES	FIXES
Operator did not follow procedure	Remind all operators to follow procedures
Did not know about procedures	Issue a "read and sign"
Inadequate procedures	Write another procedure
Operators still not following procedures	Blame the operator – Blame, Shame, Retrain
Operators STILL not following procedures	Fire the operator, "they're a bad apple"

CAPTAIN HINDSIGHT

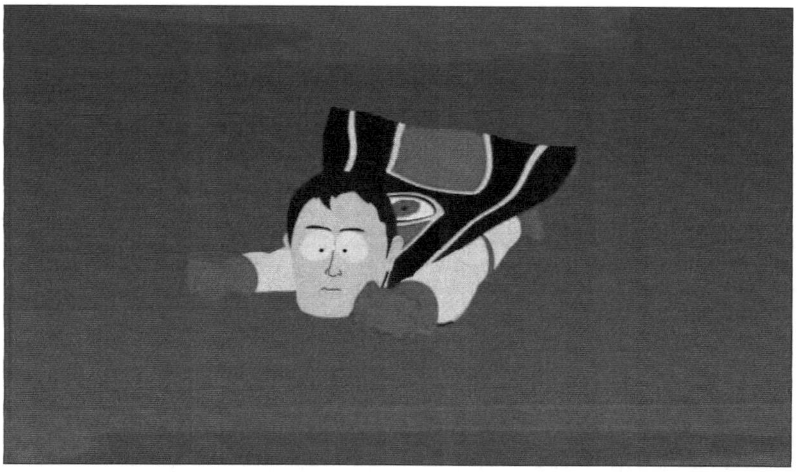

LOOKING FOR ROOT CAUSE OR CAUSES...

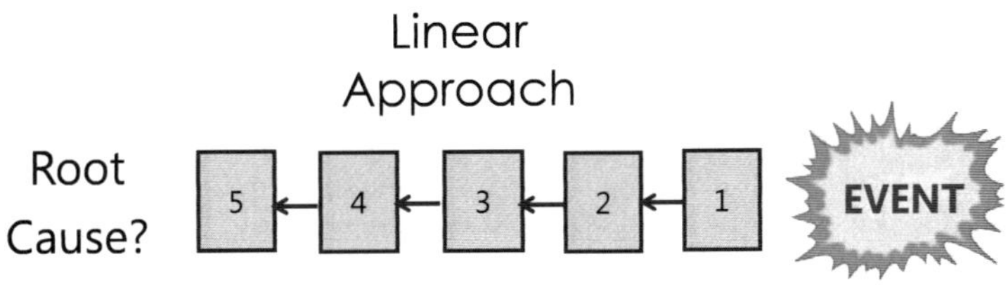

The problem is, the failure probably was not linear . . .

. . . and there almost NEVER is one root cause.

THE DOMINANT ACCIDENT MODEL

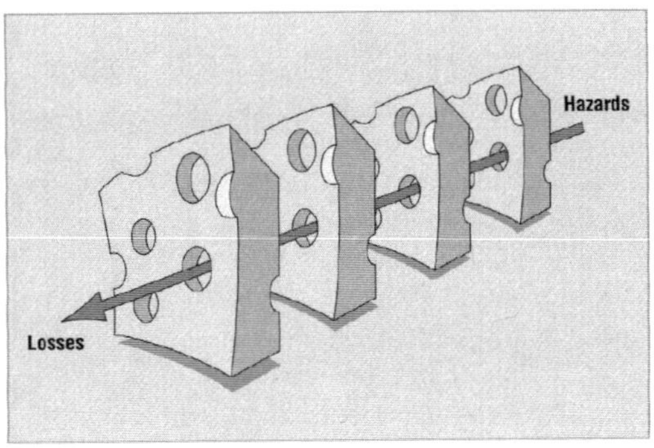

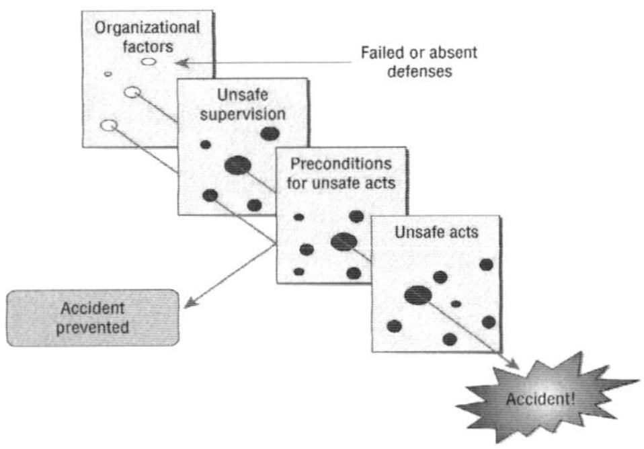

THE SWISS CHEESE FALLACY

"The SCM **does not** provide a detailed accident model or a detailed theory of how the multitude of functions and entities in a complex socio-technical system interact and depend on each other.

REASON ET AL., 2006

THE SWISS CHEESE FALLACY

"the SCM has an indisputable value as a means of communication, as a heuristic **explanatory** device.

REASON ET AL., 2006

WORK AS PLANNED vs WORK IN PRACTICE

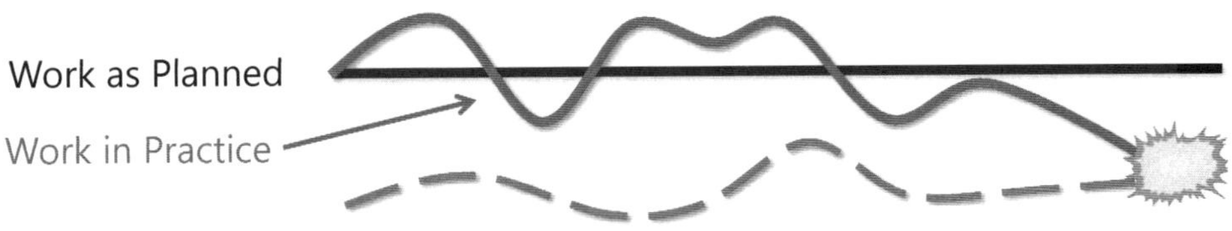

"Workers are Masters of Complex Adaptive Behaviour"

"masters of the blue line"

Conklin & Edwards, 2018

A 'MESSY' STORY

(Conklin, Edwards, Baker & Howe, 2017)

COMPLEX VS LINEAR

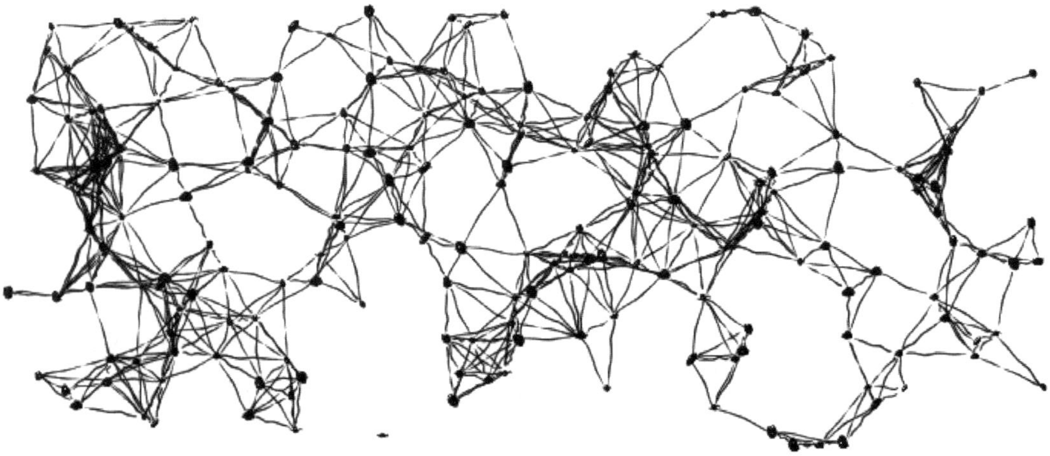

A 'MESSY' STORY

**Workers
Are Masters of
Complex Adaptive
Behavior...** (Conklin)

NORMAL ACCIDENTS?

> In our passion to understand and create an orderly universe we explain the unusual event by **invoking the usual and proclaiming it to be different**, when of course it is not.
>
> Instead, it is the **obscure, accidental, and even random concatenation of normal disorders** that produces a great even that we assume must have had great causes.

The equivalence principle...

CHARLES PERROW (1984)

SAFETY 2

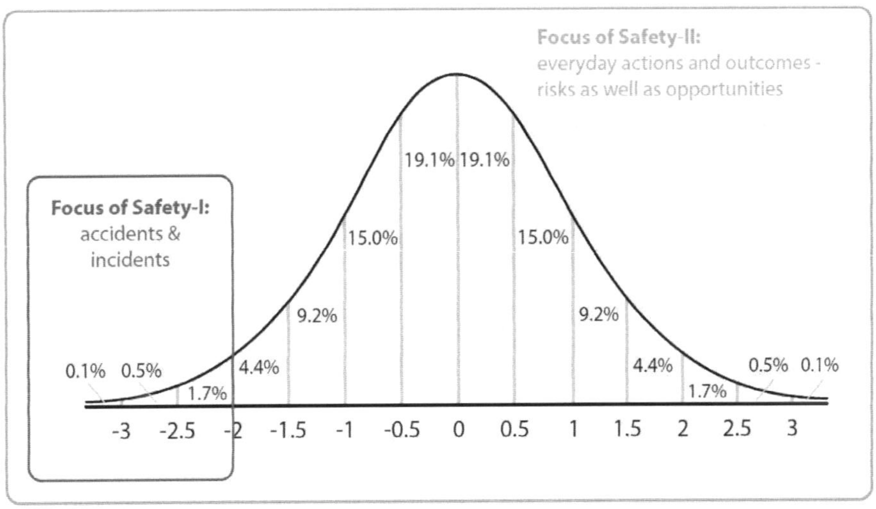

EXAMPLE QUESTIONS

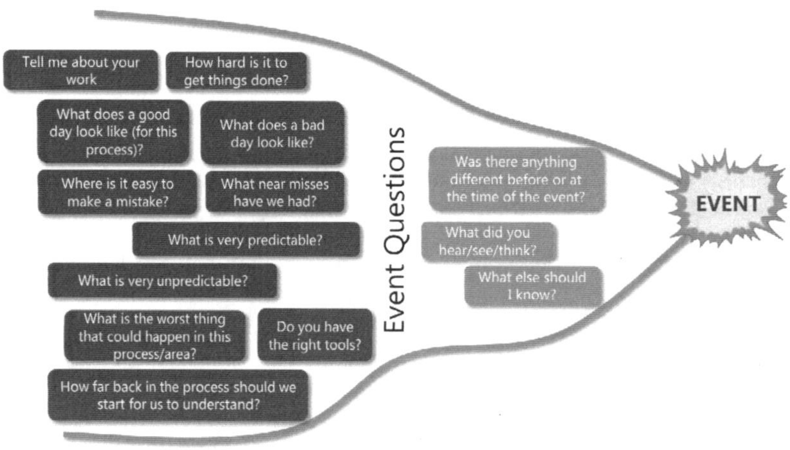

(Andrea Baker)

THE HOP PRINCIPLES

- error is normal
- blame fixes nothing
- systems drive behaviour
- learning is vital
- response matters

HOP | LAB

www.southpacinternational.com/hoplab

SOUTHPAC
aerospace

Andy Shone, CEO
Southpac Aerospace

www.southpac.biz
andy@southpac.biz

THANKYOU!

Lufthansa
Aviation Training

APATS 2019
Applying Virtual Reality into Cabin Crew Safety Training

03.09.2019 MUC N/CC-B
Birdy Raghoonundun Gunputh
Product Manager Safety Training

APATS 2019

Applying VR into Cabin Crew Safety Training

VR technology breakthrough

Implementing VR at Lufthansa Aviation Training

Lufthansa Aviation Training: re-defining tomorrow´s training

Lufthansa Aviation Training: VR tools & systems

VR: a promising technology

Lufthansa
Aviation Training

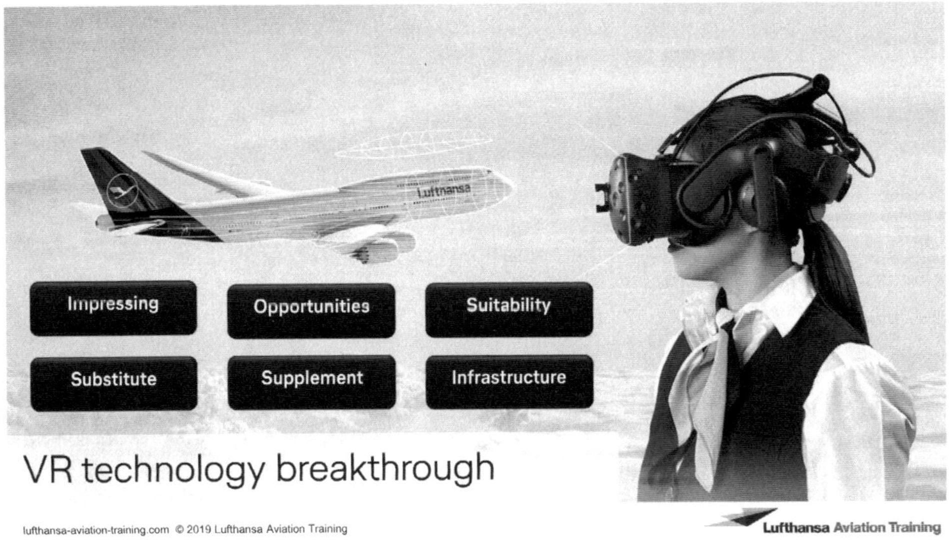
Impressing
Opportunities
Suitability
Substitute
Supplement
Infrastructure
VR technology breakthrough
lufthansa-aviation-training.com © 2019 Lufthansa Aviation Training
Lufthansa Aviation Training

Training need
Limited time
18,500 CCMs
Certified trainer
Options
Costs
Implementing VR at
Lufthansa Aviation Training
lufthansa-aviation-training.com © 2019 Lufthansa Aviation Training
Lufthansa Aviation Training

VR Training
Cabin 02
Lufthansa
Lufthansa Aviation Training:
Re-defining tomorrow´s training
lufthansa-aviation-training.com © 2019 Lufthansa Aviation Training
Lufthansa Aviation Training

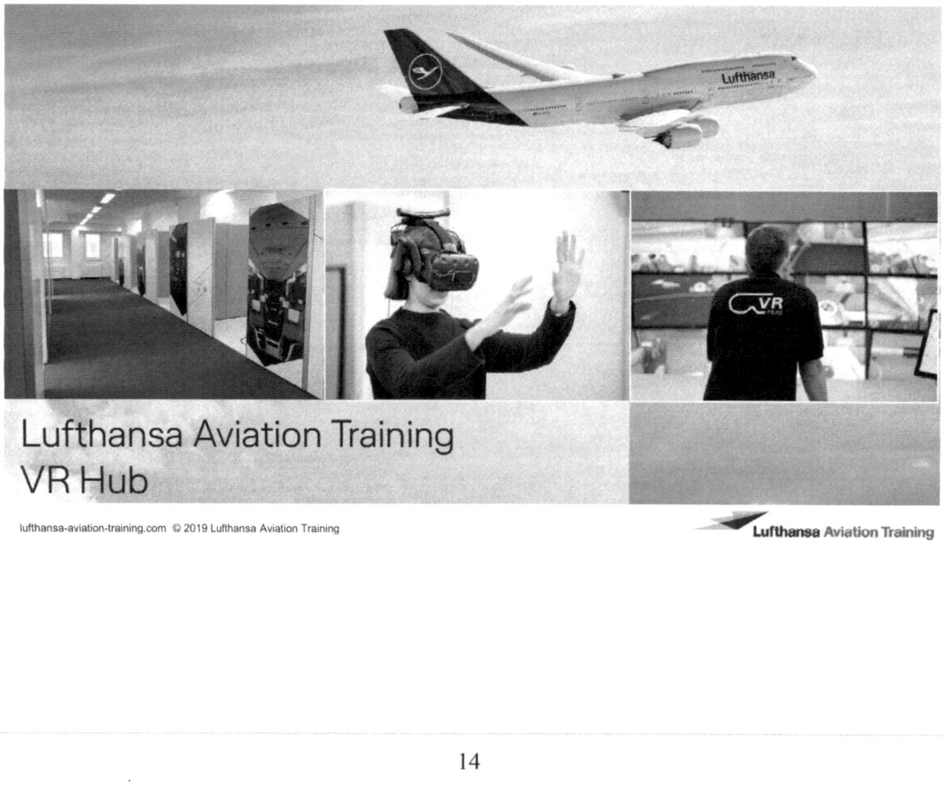
Lufthansa
VR
Lufthansa Aviation Training
VR Hub
lufthansa-aviation-training.com © 2019 Lufthansa Aviation Training
Lufthansa Aviation Training

lufthansa-aviation-training.com © 2019 Lufthansa Aviation Training

Lufthansa Aviation Training

lufthansa-aviation-training.com © 2019 Lufthansa Aviation Training

Lufthansa Aviation Training

Lufthansa Aviation Training
VR visual tracking system
lufthansa-aviation-training.com © 2019 Lufthansa Aviation Training
Lufthansa Aviation Training

Aircraft familiarization
Door procedure training
Aircraft systems
Abnormal situations
Practical checks
Equipment
lufthansa-aviation-training.com © 2019 Lufthansa Aviation Training
Lufthansa Aviation Training

Higher learning effect
Cost effective
Suitable for CBAT
Limited use
High resolution need
VR: a promising technology
lufthansa-aviation-training.com © 2019 Lufthansa Aviation Training
Lufthansa Aviation Training

Lufthansa
Aviation Training

apats2019
SINGAPORE
Keynote Address
3 September 2019
Capt. Chris Kempis
Director Flight Operations
Cathay Pacific Airways

Growth and challenges
SINGAPORE AIRLINES
JAPAN AIRLINES
JAL
Freedom of space
malaysia
© Elizabeth

Demand for air travel

CARGO

Demand for LCCs
scoot
AirAsia.com
© CNTraveller

Taking responsibility

Our greatest asset

Correction overdue...

ICAO Commercial Pilot Demand Forecasts

Region	2017 Total	2037 Total	2037 New*
Africa	11,610	28,915	23,457
Asia/Pacific	110,461	294,362	244,975
Europe	91,393	150,926	111,236
Latin American/Caribbean	24,189	48,454	36,374
Middle East	23,374	55,755	44,692
North America	98,310	141,744	93,570
World	359,337	720,156	554,304

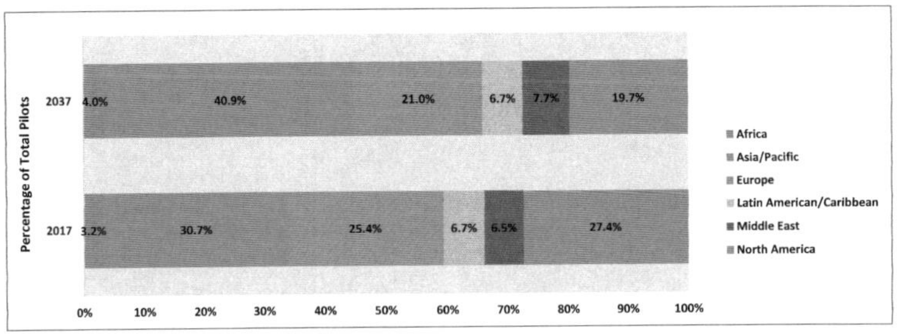

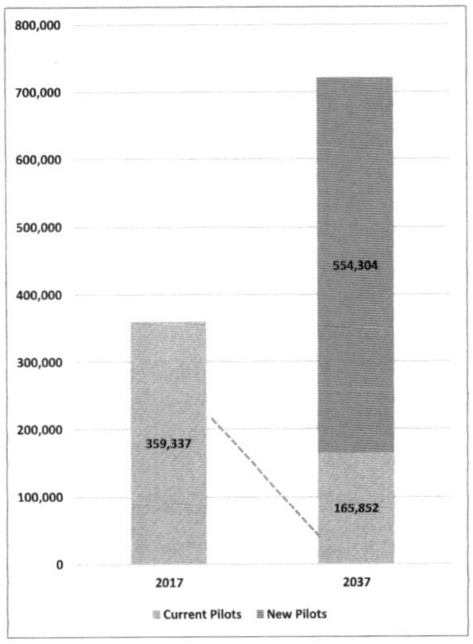

* Assumes 3% attrition rate

ICAO proprietary data, valid as of December 2018

ICAO Maintenance Technicians and Cabin Crew Demand Forecasts

Maintenance Technicians

Region	2017 Total	2037 Total	2037 New*
Africa	14,081	35,008	28,407
Asia/Pacific	132,910	354,931	295,572
Europe	114,084	181,408	132,690
Latin American/Caribbean	30,310	59,869	44,734
Middle East	24,718	59,501	47,760
North America	127,174	179,015	116,827
World	443,277	869,732	665,990

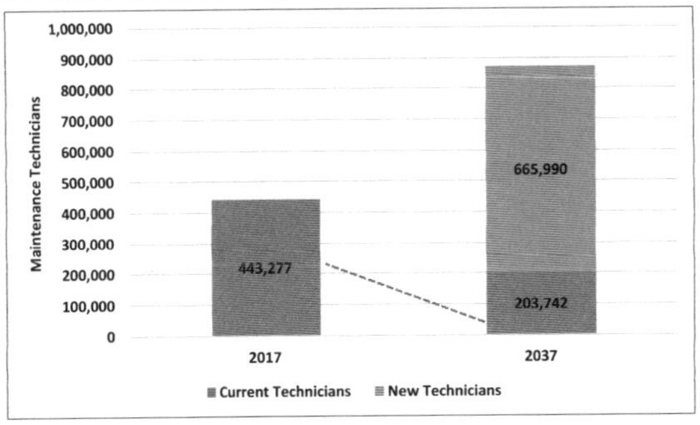

Cabin Crew

Region	2017 Total	2037 Total	2037 New*
Africa	18,660	46,239	37,452
Asia/Pacific	182,215	484,718	402,826
Europe	139,110	250,080	187,882
Latin American/Caribbean	36,080	74,644	56,719
Middle East	49,559	116,771	93,435
North America	139,540	213,096	144,864
World	565,164	1,185,548	923,178

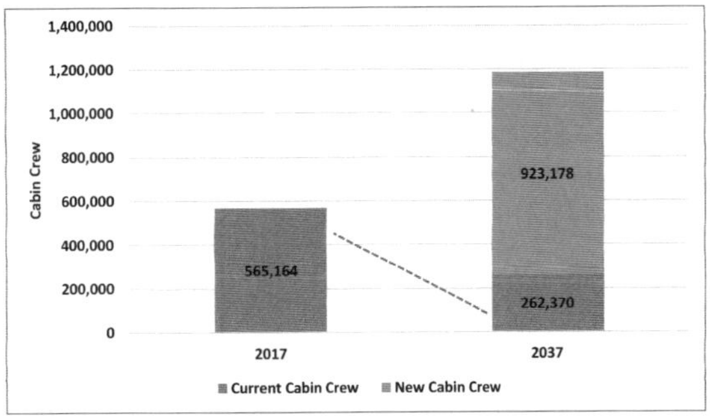

* Assumes 3% attrition rate

ICAO proprietary data, valid as of December 2018

On the flight deck

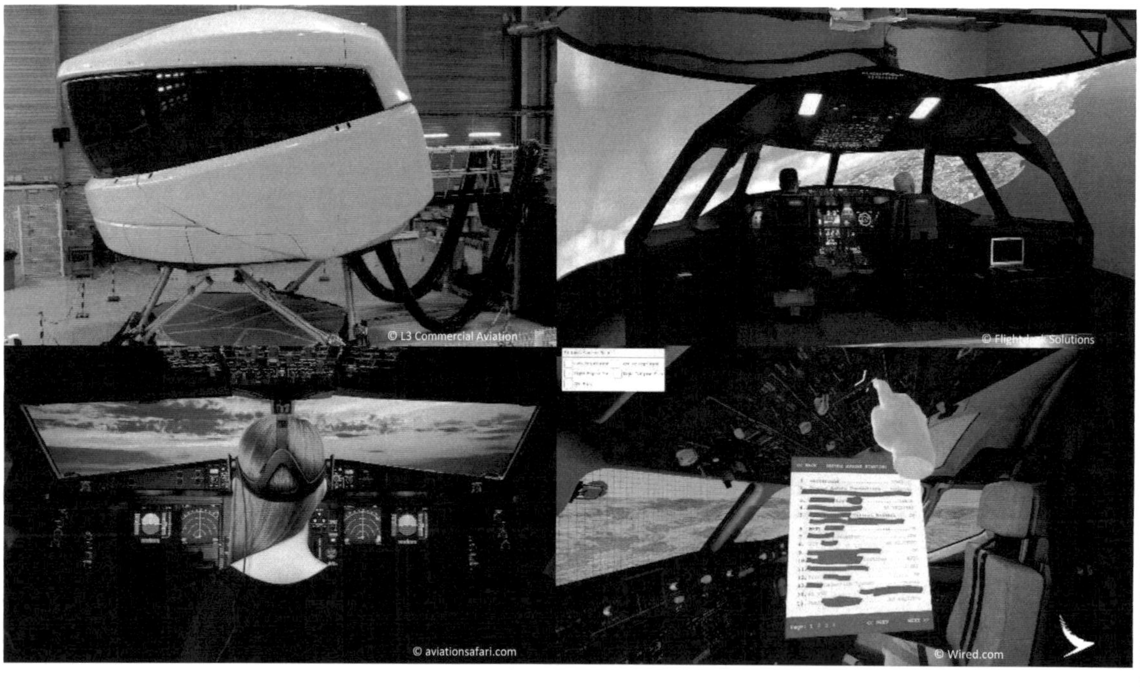
© L3 Commercial Aviation
© Flightdeck Solutions
© aviationsafari.com
© Wired.com

Demand and the next step?
A350

In the cabin

OPEN
© Bolverk XR

A clash of priorities?

In the maintenance space

© Airbus S.A.S.
© Airbus
© ESI Group

New Skillsets?

Propulsion and training

decision
strategy
ethic
vision
integrity
Leadership
contribution
communication
motivation
influence

CATHAY PACIFIC
©Andrew Hunt

"Air Rage"

A systematic review of research on Disruptive Airline Passenger Behaviour

APATS Singapore 2019

Dr Doug Drury

University of South Australia

Media

- **Air rage woman who stormed cockpit on Jet2 flight yelling 'I'll kill you all' slapped with $150k airline bill -** ***News.com.au***
- **Man 'slaps woman' as air rage brawl erupts on Turkish Airlines holiday flight from Istanbul to Russia –** ***The Sun***
- **Air-rage drunk belted boy, 8, on head with bottle and spat in woman's face on Virgin flight –** ***The Scottish Daily Record***

University of South Australia

Media

- **Air rage: how it differs in West and East, and why Chinese passengers are in a league of their own –** ***South China Morning Post***

- **Two 'violent' passengers kicked off easyJet flight from Gatwick to Gibraltar after being 'aggressive' towards crew –** ***The Sun***

- **Brits named most unruly airline passengers by Spanish authorities –** ***The Mirror.com.uk***

University of South Australia

Media

- **Pilot takes down passenger who pushed flight attendant**

- **Male passenger sparks complete disgust as he urinates in his plane seat –** ***Express.co.uk***

- **Passenger Accused of Defecating on Airliner Pleads Guilty To Making A Threat –** ***AP News***

University of South Australia

Disruptive Passengers

Unruly passengers are those whose demeanour, behaviour or failure to comply with cabin crew directions present a threat to the safety or security of the aircraft and/or those on board. (Goldsmid et al., 2016)

University of
South Australia

Statistics

- Every three hours there is a disruptive passenger event within the European Union - *EASA*
- 1 incident in 1,424 flights in 2017 - *IATA*
- 20,000 airplanes in the air during peak travel – *FlightRadar24*
- 1 incident in 14 flights
- 49% are non-compliance with safety regulations - *IATA*
- 27% report alcohol involved - *IATA*

University of
South Australia

Disruptive Behaviour Cause?

- **Travel Stress**
  - Begins before arriving at the airport
- **Alcohol/Drugs**
  - Widely reported in print and social media
- **Class Anger**
  - Business vs Economy
- **Airplane Ergonomics**
  - Economy seat size

University of South Australia

Disruptive Passenger Management

- **What are the "Best Practices"**
  - Each airline culture is different
  - Passenger culture changes
- **Where are the Gaps?**
  - Training – Is it effective?
  - Organizational Support – Is it enough?
  - Regulatory – Does it need to change?
  - Air Law – Does it need to change to deter others?

University of South Australia

Impact on Cabin Crew

- **Effectiveness/Confidence in Containing Event**
  - Self-defence training
  - Crisis negotiation training
- **Recovery**
  - PTSD Treatment
  - R U OK?
- **What Do You Need?**

University of South Australia

How Can We Help?

- **Incident Data Analysis**
  - Airline data from incident report
  - Deidentified
- **"Best Practices" Analysis**
  - What is working, what is not
- **Training Methodologies**
  - Use of technology to develop integrated training

University of South Australia

Contact Information

Dr Doug Drury
Program Director
Aviation
University of South Australia
doug.drury@unisa.edu.au
61+ 0883023945

University of
South Australia

INCLUSION MOVES

INCLUSION MOVES SOCIETY FORWARD

inclusionmoves.com.au

Training to Handle Disabled Passengers

Balancing Customer Service, Risk and a Human Rights Approach

Geoff Trappett OAM - Founding Director of Inclusion Moves

Simple. Straightforward. Uncomplicated social inclusion means a healthy society for all.

Scene Setting

The interplay between customer service, risk and human rights is never more evident than in the customer service provided to passengers with a disability.

In this session we will explore these intersections, how organisational and societal culture plays its part and crucially how to ensure your business enables a human rights focus to shine through in your customer service delivery model.

Disabled people are becoming more and more full citizens of their communities. This brings opportunity and challenge to those communities. We must ensure we are up to this challenge.

INCLUSION MOVES

INCLUSION MOVES SOCIETY FORWARD

What is the regulatory environment we are presented with?

- UN - Convention on the Rights of Persons with Disabilities - Guiding document with the purpose of pursuing the rights of disabled people
- Regulation (EC) 1107/2006 - Europe - Speaks to the rights of disabled passengers within the EU
- Air Carrier Access Amendments Act - America - Speaks to the rights of disabled passengers in the US. Also introduces mandatory reporting for loss/damage of mobility aids.
- Montreal Convention - International treaty in this context speaking to the liability for costs for loss/damage to mobility aids
- Civil Aviation Order 20.16.3. and 20.11. CASA regulations ensuring consistency in service delivery for disabled passengers in Australian context.

This is not intended to be a thorough list. Much the opposite it is meant to show the complications of customer service in a heavily regulated environment.

This uncertainty brings a number of questions:

How does this heavy regulatory burden sit alongside what disabled passengers wanting from a customer experience?

Are we aware what the customer experience they are wanting is? Are we engaged in our local disabled communities to discuss these matters?

Do we have the background systems in place to ensure we are able to provide a good customer experience?

How do we map and improve the customer experience for able bodied passengers? When was the last time this was done for the disabled cohort? What if anything would need to change in your customer experience mapping frameworks to achieve this?

Do your staff hold the skills, culture and emotional intelligence to be disability confident? Does your leadership inspire this?

Developing a framework for answering these vital questions authentically and with diligence

Equality: the state of being equal, especially in status, rights, or opportunities.

Equity: the quality of being fair and impartial.

Liberation: the action of setting someone free from imprisonment, slavery, or oppression; release.

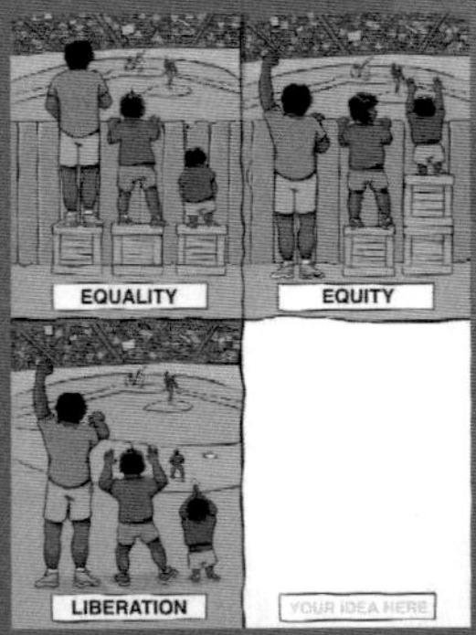

How are you working to not only bring equality and equity to your customer service but crucially working to break down barriers and bring liberation to ALL your customers?

You must develop a **culture of inclusion**

IM INCLUSION MOVES

INCLUSION MOVES SOCIETY FORWARD

Developing a culture of inclusion - disability confidence

- Must be authentic
- Must be disabled community led
- Must be led from the top
- Must have the systems in place to ensure good policy and procedures can be enacted
- Must be long term/adaptable
- Must be demonstatable and measurable

Nothing should happen about a person with a disability without a person with a disability being central to the conversation. Our purpose is to give people with a disability the unhindered opportunity to ensure they are able to enact their full citizenship rights like any other. Equal citizenship rights means accessing the Same jobs – Same venues – Same services – Same social opportunities.

IM INCLUSION MOVES

INCLUSION MOVES SOCIETY FORWARD

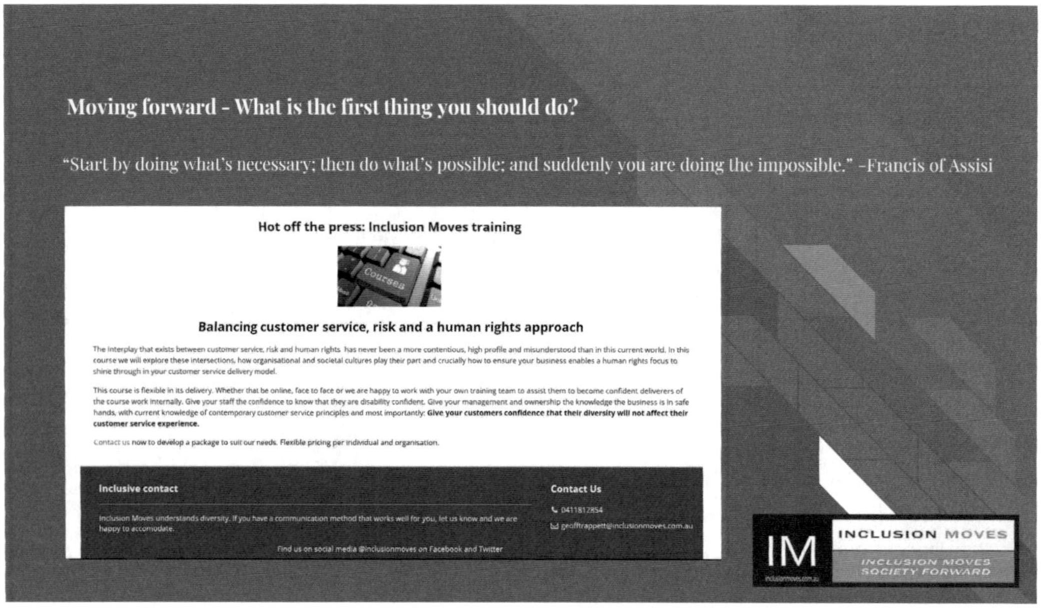
Moving forward - What is the first thing you should do?
"Start by doing what's necessary; then do what's possible; and suddenly you are doing the impossible." -Francis of Assisi
Hot off the press: Inclusion Moves training
Courses
Balancing customer service, risk and a human rights approach
The interplay that exists between customer service, risk and human rights has never been a more contentious, high profile and misunderstood than in this current world. In this course we will explore these intersections, how organisational and societal cultures play their part and crucially how to ensure your business enables a human rights focus to shine through in your customer service delivery model.
This course is flexible in its delivery. Whether that be online, face to face or we are happy to work with your own training team to assist them to become confident deliverers of the course work internally. Give your staff the confidence to know that they are disability confident. Give your management and ownership the knowledge the business is in safe hands, with current knowledge of contemporary customer service principles and most importantly: Give your customers confidence that their diversity will not affect their customer service experience.
Contact us now to develop a package to suit our needs. Flexible pricing per individual and organisation.
Inclusive contact
Inclusion Moves understands diversity. If you have a communication method that works well for you, let us know and we are happy to accomodate.
Contact Us
0411812854
geofftrappett@inclusionmoves.com.au
Find us on social media @inclusionmoves on Facebook and Twitter
IM
inclusionmoves.com.au
INCLUSION MOVES
INCLUSION MOVES SOCIETY FORWARD

Professional Sleeping For Cabin Crew

Professional Sleeping For Cabin Crew

Professional Sleeping
For Cabin Crew

Professional Sleeping
For Cabin Crew

Professional Sleeping
For Cabin Crew

Alcohol

Alcohol

Alcohol

Bed Bugs

Bed Bugs

Bed Bugs

Body Clock

Body Clock

Body Clock

Coffee

Coffee

Coffee

Illegal Drugs

Illegal Drugs

Illegal Drugs

Legal Drugs

Legal Drugs

Legal Drugs

Fitness

Fitness

Fitness

iPhones, Audiobooks & Apps

iPhones, Audiobooks & Apps

iPhones, Audiobooks & Apps

Massages

Massages

Massages

Mosquitoes

Mosquitoes

Mosquitoes

Natural Remedies

Natural Remedies

Natural Remedies

Nicotine

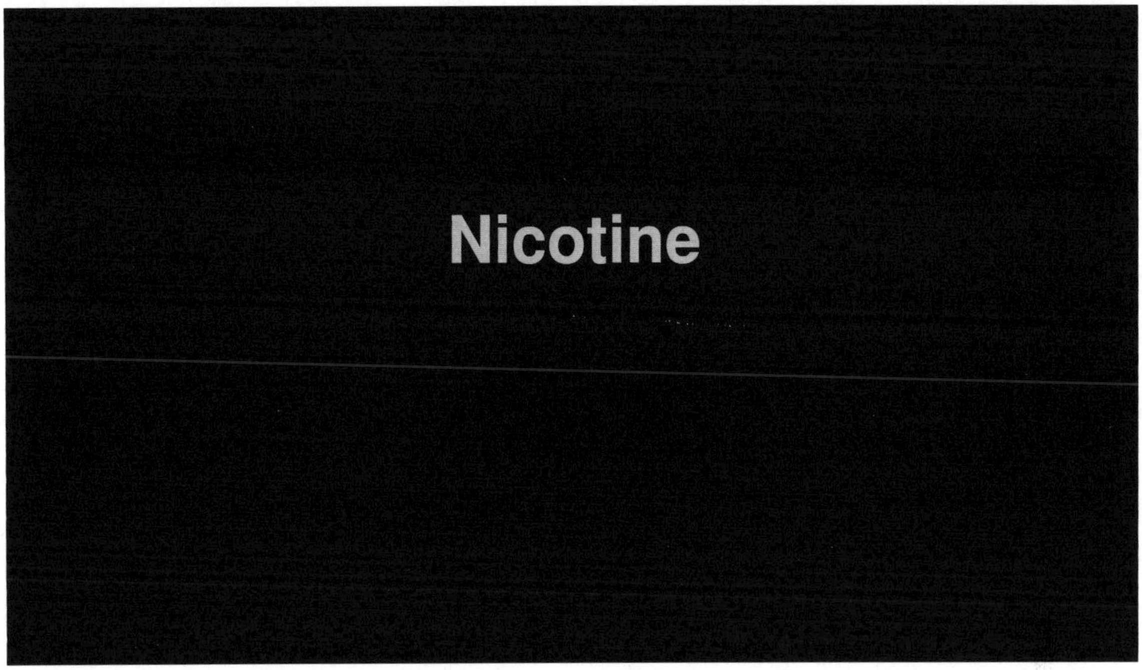
Nicotine

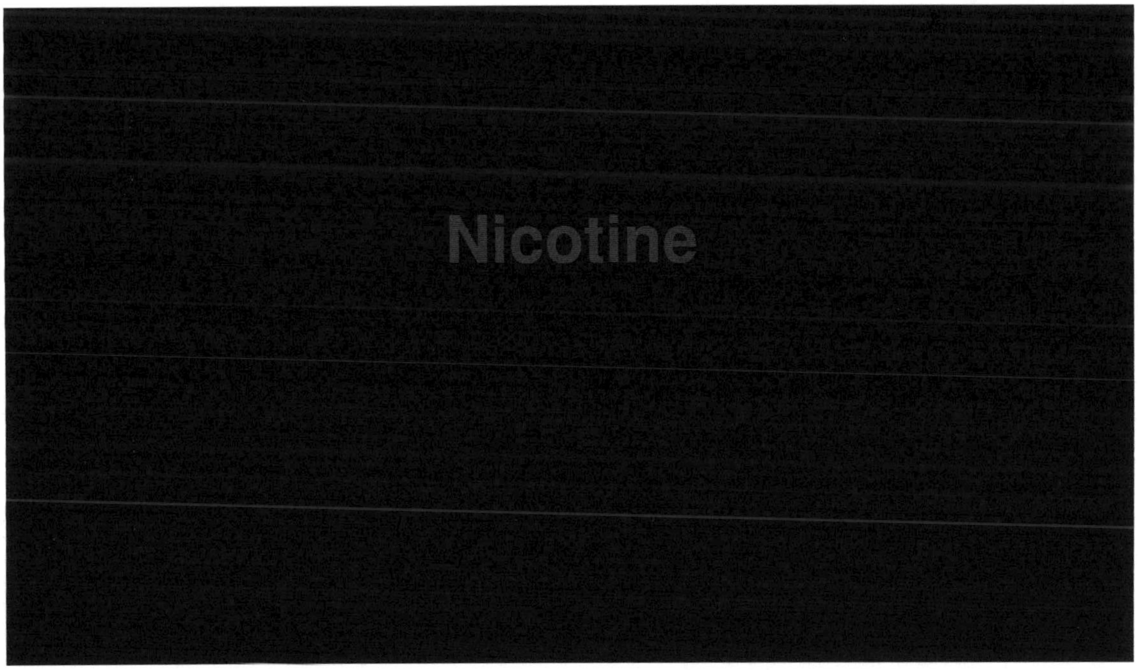
Nicotine

Rosters & Diaries

Rosters & Diaries

Rosters & Diaries

Snoring

Snoring

Snoring

Sleep Apnea

Sleep Apnea

Sleep Apnea

Stress

Stress

Stress

Sun - Too Much

Sun - Too Much

Sun - Too Much

Sun - Not Enough

Sun - Not Enough

Sun - Not Enough

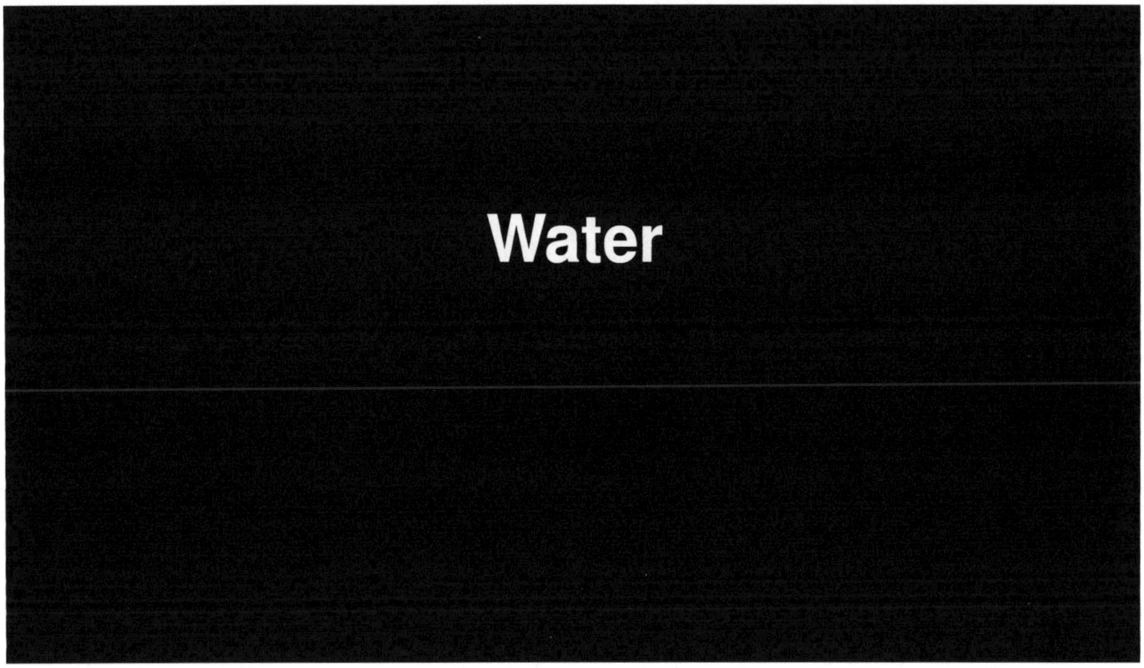
Water

Water

Water

Weight & Sex

Weight & Sex

Weight & Sex

Sleep Preparation

Sleep Preparation

Sleep Preparation

Room, Bed, & Position

Room, Bed, & Position

Room, Bed, & Position

Self-Hypnosis & Relaxation Techniques

Self-Hypnosis & Relaxation Techniques

Self-Hypnosis & Relaxation Techniques

Wake Up Naturally

Wake Up Naturally

Wake Up Naturally

Fake It Til You
Make it!

Fake It Til You
Make it!

Fake It Til You
Make it!

Crew Rest Coffin

Crew Rest Coffin

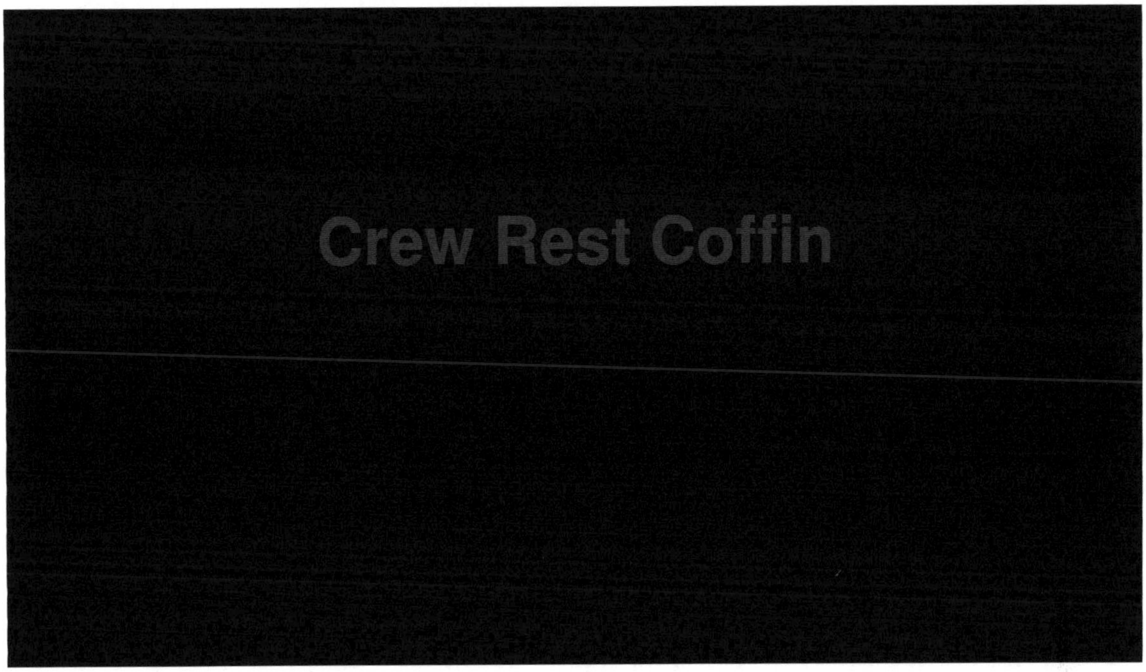
Crew Rest Coffin

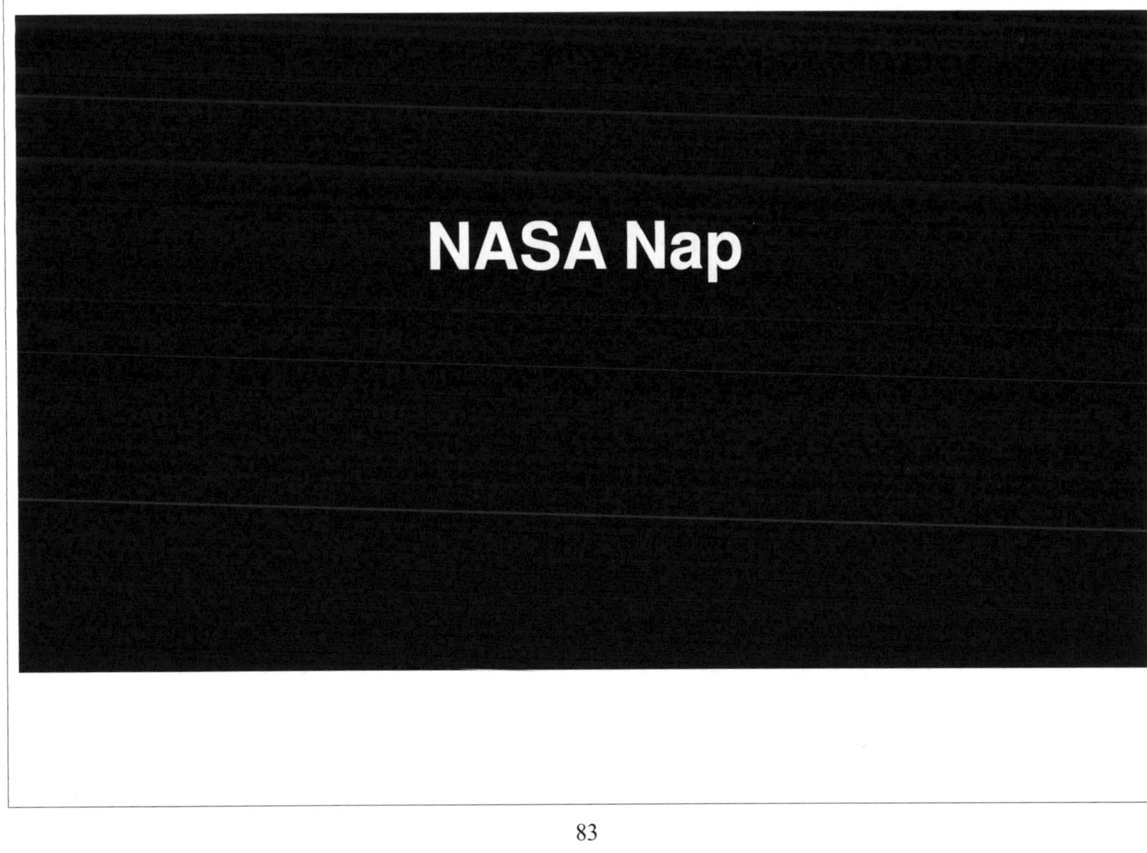
NASA Nap

NASA Nap

NASA Nap

Layover Destination

Layover Destination

Layover Destination

Jet Lag

Jet Lag

Jet Lag

Expat Lifestyle

Expat Lifestyle

Expat Lifestyle

A Long Flight

A Long Flight

A Long Flight

Sleep Professionally

Sleep Professionally

Sleep Professionally

1. Your #1 Priority

1. Your #1 Priority

1. Your #1 Priority

2. Alarms, Earplugs, Eyeshades

2. Alarms, Earplugs, Eyeshades

2. Alarms, Earplugs, Eyeshades

3. No Electricity

3. No Electricity

3. No Electricity

james@jamesnixon.com

AIRBUS A380-800
james@jamesnixon.com

独断-주장-自己主張-sự quyết đoán

Assertiveness in the Asian Culture

Joey Anca
Human Risk Manager, Metro Trains Melbourne

17 September 2019

METRO

Joey Anca

Human Risk Manager
Operations

METRO

1986-2017

Philippine Airlines | Ansett Australia | ICAO • OACI • ИКАО | Singapore Airlines | Transport Sydney Trains | Jetstar | Swinburne University of Technology

- Multimodal Safety Management Systems & Human Factors
- Crew Resource Management, 2010 (NASA-University of Texas at Austin)
- Crew Resource Management, 2018 (NASA-US Federal Aviation Administration)

Assertiveness

独断 (Dúduàn)
주장 (Jujang)
自己主張 (Jiko shuchō)
sự quyết đoán

Assertiveness involves three dimensions: **Active Listening, Speaking Up and Performing an Action**

Outline

1. Socio-Political/ Socio-Cultural contemporary foundations of assertiveness in Asian culture
2. Small Group Activity on Assertion
3. Aviation Example of Non-Assertiveness
4. Reviewing Culture, CRM & Error

17 September 2019 | 3

Socio-Political Contexts

60.09%
of the World under Dictatorship is in Asia

Population under Dictatorship
World=2.626B
Asia=1.601B

60.09% Asian

Dictatorship Countries Population. (2019-08-15). Retrieved 2019-09-01, from http://worldpopulationreview.com/countries/dictatorship-countries/

17 September 2019 | 4

Socio-Cultural Context

Socio-Cultural Context

Influences of Socio-Cultural/Political Contexts on Asian Disposition to *Assertiveness*

- Strong team orientation; bonded together to protect shared Values
- Protect collective rights and principles
- Deep sense of Compassion for other team members
- Facilitates operation during emergencies (power-distance) requiring command and control

- High Power-Distance can discourage information-sharing
- Promoting more assertion, as "assertiveness training" for junior members to seniors may not work
- Asking for group input may be difficult

Stereotyping Asian Docility (Low-Assertion)

South China Post (2014, Jingan Young):

- Cabin crew uniforms figured out in a dispute (2014, CX) where the Union noted CX cabin crew uniforms were "too revealing" in that these were "too tight, too short", which have contributed to an increase in sexual harassment occurrences especially with members of the airline's frequent flyer (Marco Polo) Club

Question: Is the uniform really to blame?

There is a more troubling issue at work here which have to do with **perceptions of Asian women as "submissive and docile"** largely contribute to the unwanted attention and treatment they receive in the workplace.
In a now defunct Cathay campaign features a photograph of a flight attendant who was quoted as saying "I love helping people even before they ask".

Assertion Activity

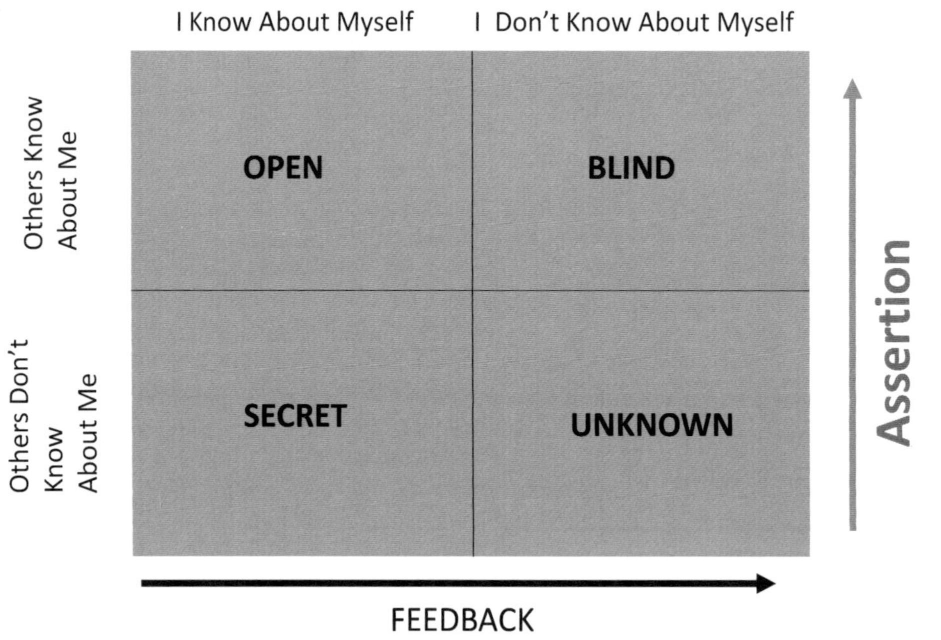
I Know About Myself
I Don't Know About Myself
Others Know About Me
Others Don't Know About Me
OPEN
BLIND
SECRET
UNKNOWN
Assertion
FEEDBACK

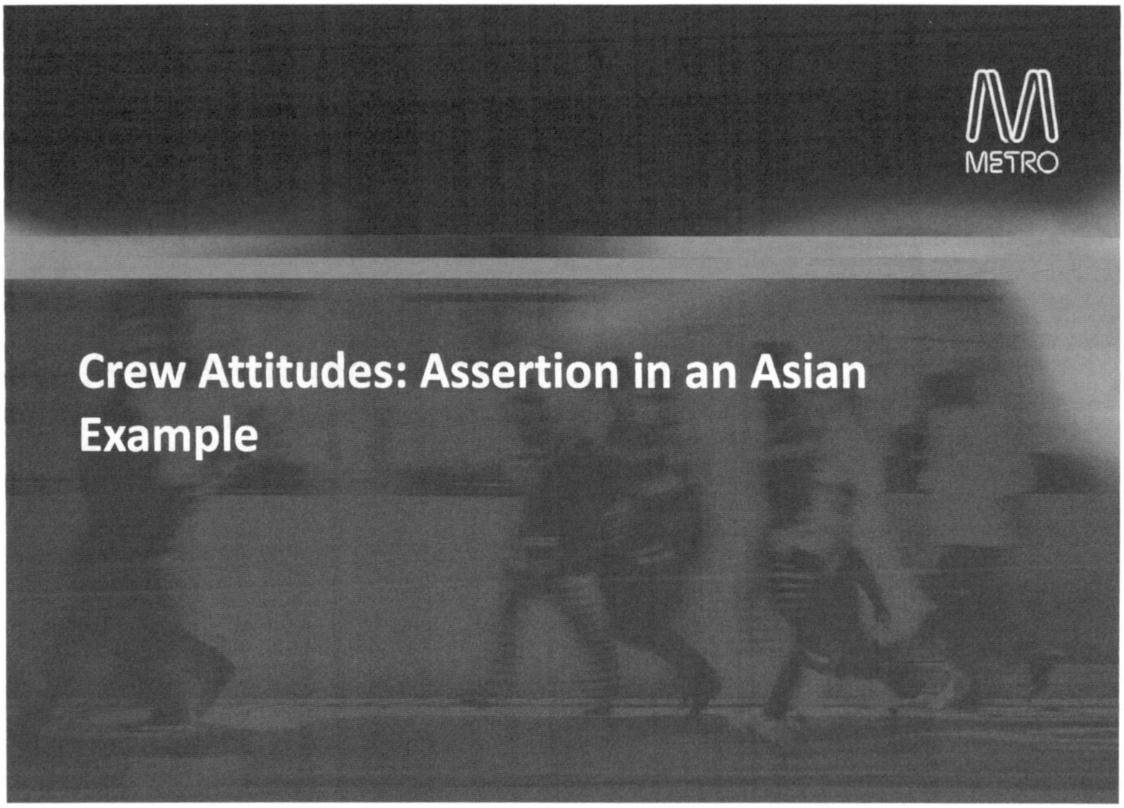
METRO
Crew Attitudes: Assertion in an Asian Example

Excerpt from the (Cockpit) Voice Recorder, CVR

"I knew this was the first time the captain was flying the Hongkong IGS Visual Step Down approach. My student Flight Engineer saw the anti-skid fail light come on. I told him to just switch it (all) off. At about 300 ft AGL, I could see the captain having a hard time at the controls during the approach..."

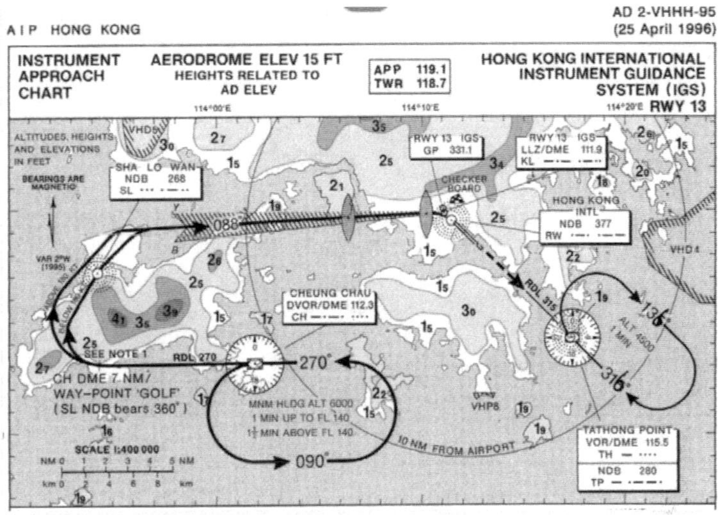

B747-200 Flight Engineer's Panel

Accident Findings

- The Aircraft had no Anti Skid Protection at all
- Twelve (12) tyres burst after touchdown
- Aircraft was disabled on RWY13
- All incoming flights were diverted to alternate airports
- Organisational & Political: Bad publicity for the airline

Crew Demographics

Captain

- 13,000 Total Flying Time
- First Solo to HKG after 2 Route Checks
- Previously B737-300 Captain flying domestic

First Officer

- 1,100 Total Flying Time; mostly spent in B747-200 as First Officer
- More Senior (in years of service) than the Captain

Check/Instructor Engineer

- New as a Check/Instructor Engineer
- Air Force background (Colonel)

(Trainee) Flight Engineer

- This flight was to be his Release Check as a Flight Engineer
- He was an Air Force Captain before joining the airline

The US$3.4 Million Question:

Why did the Trainee Flight Engineer obey an erroneous command from the Check/Instructor Engineer when he knew it was the wrong procedure to shut off all hydraulics on the Hongkong IGS Approach?

Exploring Cultural Solutions

Professor Geert Hofstede (1980)

88,000 Respondents from a large multinational corporation

- ***High Power-Distance Index*** (leaders are expected to be direct & decisive; followers are expected to obey
- ***Low Individualism Index*** (collectivist culture where membership in a group implies moral and personal commitment to the group)

My Investigation Findings and Assertion

This is a Cultural and Assertion Problem:

- The Trainee Engineer actively listened, did not speak out, and complied with the command from his Instructor Engineer even if he knew **it was the wrong thing to do;**
- The Instructor Engineer was of a higher rank (Colonel); he was only a Captain;
- Obeying a Command was a correct (optimizing) response rather than follow training standard.

Collectivist vs. Individualist Cultures

Individualist Culture:

- ***INDIVIDUAL*** Rights are important;
- The political system is founded on the belief that people will pursue their own self-interest and the aggregate effects of those individual pursuits, properly regulated, will culminate to the common good.

What is S.I.R.?

- Where open and free communication may be the feature of the Western Anglo Saxon Flightdeck with the utmost concern concern of arriving at a Social Product (conclusions, outcomes, agreeing to disagree etc.)
- the Collectivist Asian Team places highest importance instead, on the Social Process (tact, cautious words, overly willing to coordinate, keeping the peace etc.)

S.I.R. at HKG IGS Approach

S.I.R = "**S**mooth **I**nterpersonal **R**elationships"

- The Captain was having a hard time on the approach...(High Power-Distance)
- Anti-Skid Fail Light ON...("I am being sensitive to the situation and need to maintain trouble-free and smooth relationships)
- Obeying an erroneous Command...(Being agreeable even under difficult circumstances)

Adapted from Hollnsteiner, M & Kaut, P. (1978). Smooth Interpersonal Relationships in the Philippine Setting.

What the Airline did to Explore Cultural Solutions

- Find out the Cultural "Precursors" to Error, and not just the Individual Causal Factor to the incident;
- Survey the crew—there will be systemic (and hidden) cultural reasons;
- Engage the line pilots—they know what is happening out there;
- Establish crew resource management with cabin crew, pilots, engineers and ramp staff.

Exploring Cultural Solutions

NASA/UT/FAA Flight Management Attitudes Questionnaire examined three overarching clusters of flight management attitudes:

- Command Responsibility
- Communication & Coordination
- Recognition of Stressor Effects

Kanki, B., Helmreich, R.L. & Anca, J. (2007). Crew Resource Management, Academic Press: San Diego.

The same error, different consequences...that's why the Context of an Incident is important.

Hayward, B. & Paries, J. (2000). Concept and Graphic Dedale Asia Pacific.

Investigating the RMS Titanic

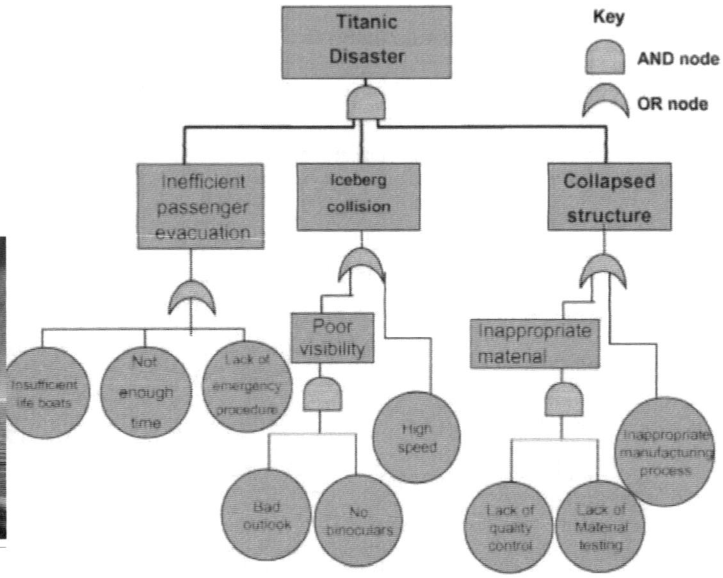

Integrated Investigations

- Nautical Advisor- not right to impose safety regulations, because maritime was the safest mode of travel
- Captain Smith ignored iceberg warnings
- Lookout had not binoculars

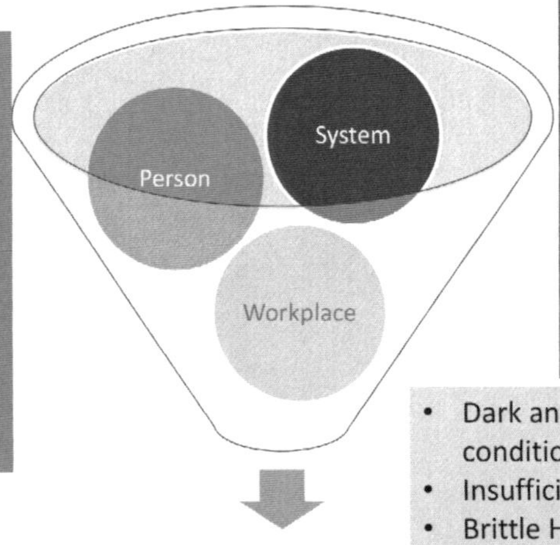

- Shortage of skilled labour
- No emergency evacuation training
- No emergency drills
- Rush to meet record voyage times—excessive speed

- Dark and poor weather conditions
- Insufficient lifeboats
- Brittle Hull Steel
- Low-grade rivets

Psychosocial Risk Management 16-18 Sept 2019, Brisbane Queensland 27

"It's Not A Swiss Cheese": It is about Overlapping Sleeves of Protection that you MUST ask your CEO

- Maturing Culture: are your **Values and Beliefs Compatible?**
- Do you **Understand your RISKS?**
- Vulnerability: Can you Predict your Next Accident?

Increasing Protection

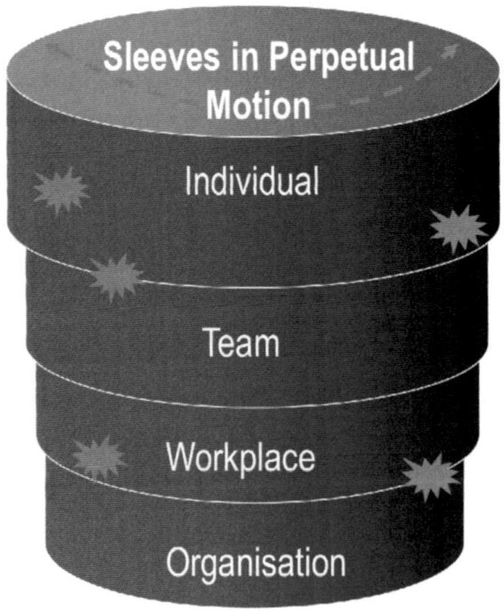

Anca, J (1993). Why Prayers bring few miracles. PAL Report. Manila:1993.

Increasing Vulnerability

- Staff Loyalty: Do we desire **Team Success or Individual Wins**?
- Are Our Leaders more concerned about KPIs and Self-Preservation or about **Safety & the Common Good**?
- Safety Information: is it **Meaningful to a few or is it Owned by ALL?**

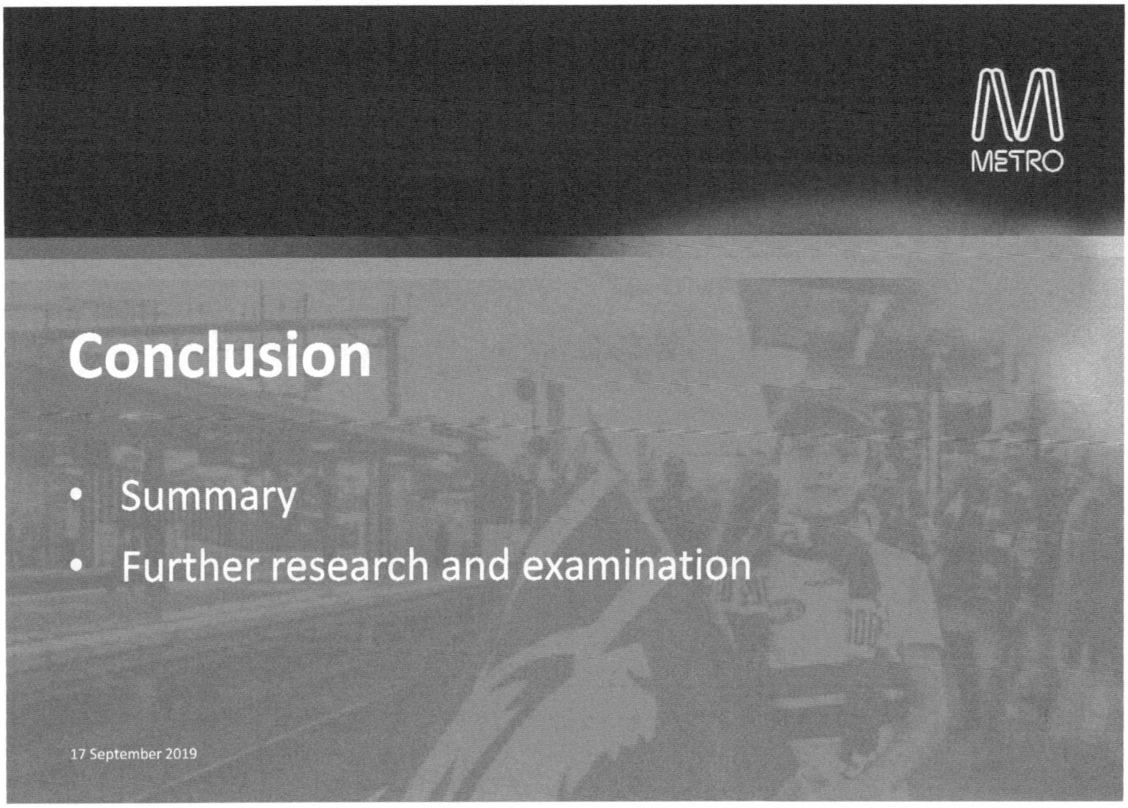

People: The Character of Action

People will only defend the organisation if it feels that the more basic concerns are addressed.

Though commercial goals will be achieved, dealing with catastrophes will be elusive

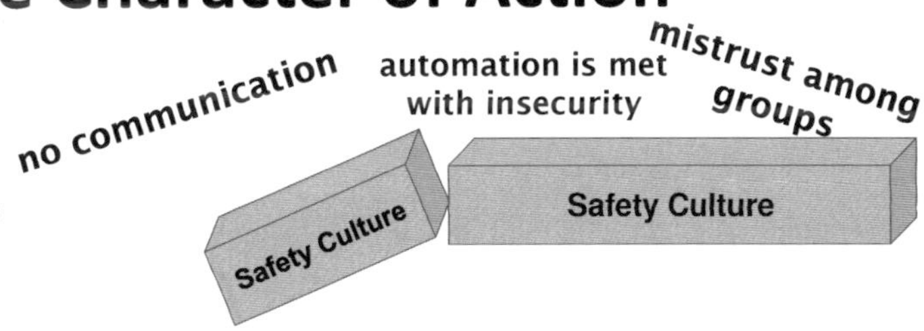

PEOPLE	SYSTEM	RISK
The Who	*The How*	The What IF
***Character** of Action*	***Process** of Action*	***Control** of Action*

What then appears is a situation where safety culture is broken and not integral to the way things are done.

Next Generation fire training

by Robin Pijnaker

Managing Director

Flame Aviation

3rd of September 2019

FLAME AVIATION 1

Introduction

Traditional fire training: classroom instruction + practical training.

Now, there are two developments:

1. A higher competency level is required for fire training
2. New technology is available which can increase the effectiveness of the training

FLAME AVIATION 2

Questions

- Is the traditional curriculum still effective?
- Can we complement stand-up instruction with CBT?
- Can we use VR?
- Can we standardize our training?
- Can performance be measured?

1. FIRE TRAINING CURRICULUM

TRADITIONAL CURRICULUM

Classroom

FLAME AVIATION

5

TRADITIONAL CURRICULUM

Classroom

1. Knowledge transfer
2. Introduction to equipment

Improvements?
- Efficiency
- Flexibility

FLAME AVIATION

6

TRADITIONAL CURRICULUM

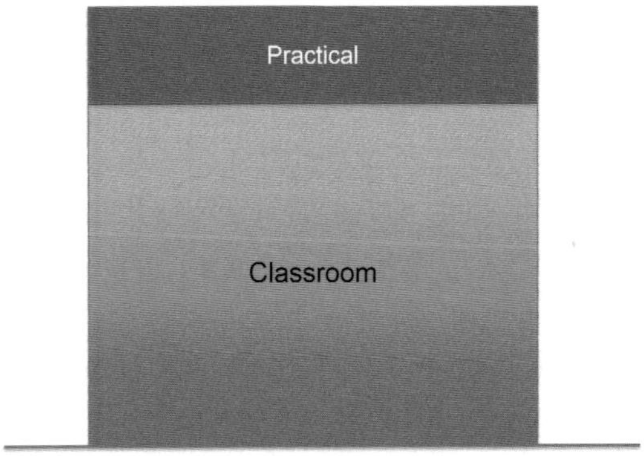

FLAME AVIATION

7

TRADITIONAL CURRICULUM

1. Exposure to real fire
2. Handling equipment

Improvements:
- Realistic (equipment, environment, etc)
- Procedures

FLAME AVIATION

8

NEW CURRICULUM

CBT

1. Efficient knowledge transfer
2. Flexible
3. Exams and certificates

NEW CURRICULUM

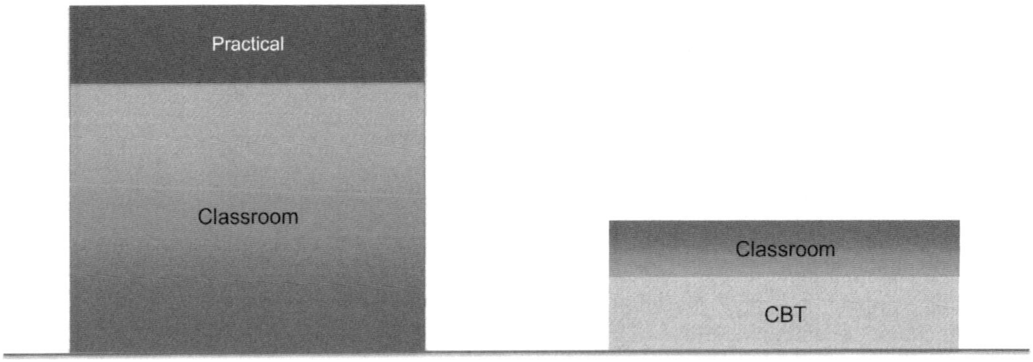

FLAME AVIATION 11

NEW CURRICULUM

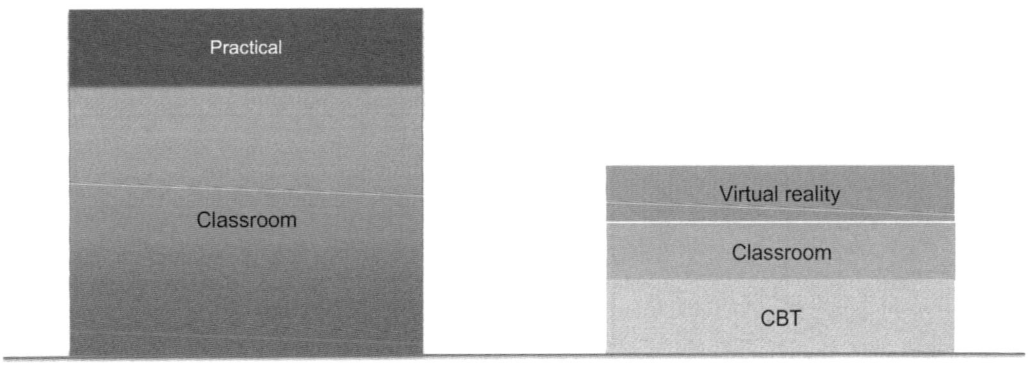

FLAME AVIATION 12

VIRTUAL REALITY

1. Preparation for practical training
2. Ideal for training procedures

NEW CURRICULUM

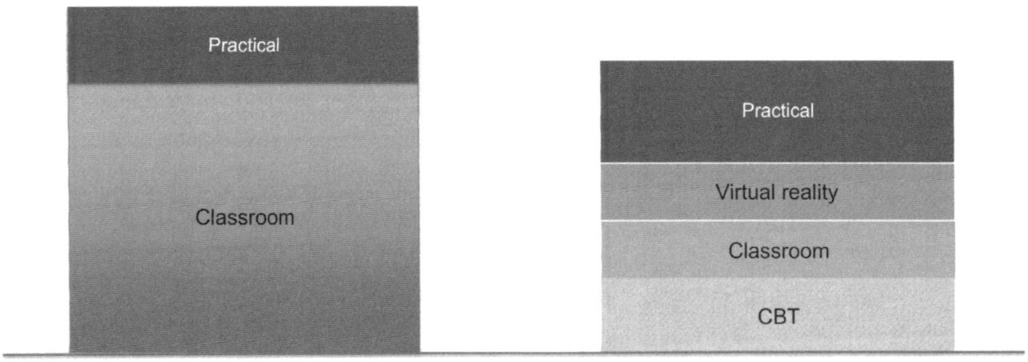

PRACTICAL TRAINING

1. Realistic training environment and equipment
2. Exposed to fire, smoke, heat, smell and sound
3. Pre-programmed scenarios

2. STANDARDIZATION OF TRAINING

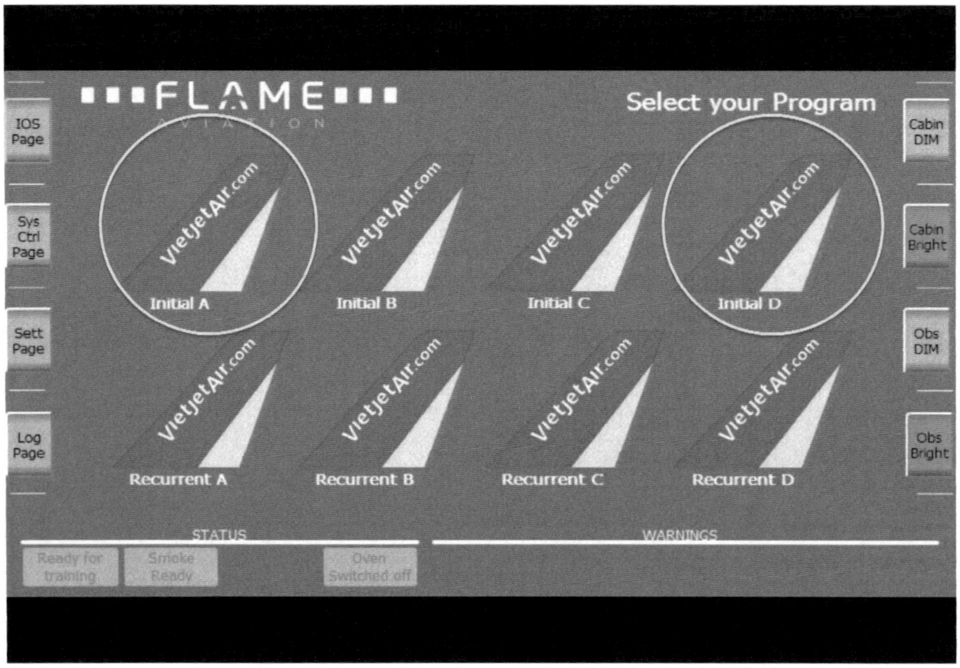

Initial A parameters settings

✓ Lighting:	100% (bright)
✓ Smoke density	15% (low)
✓ Cabin panic sound	OFF
✓ Burning smell	OFF
✓ Difficulty level	Level 1 (easy)
✓ Total training time	2,5 minutes (long)
✓ Relight time	12 seconds (long)

FLAME AVIATION

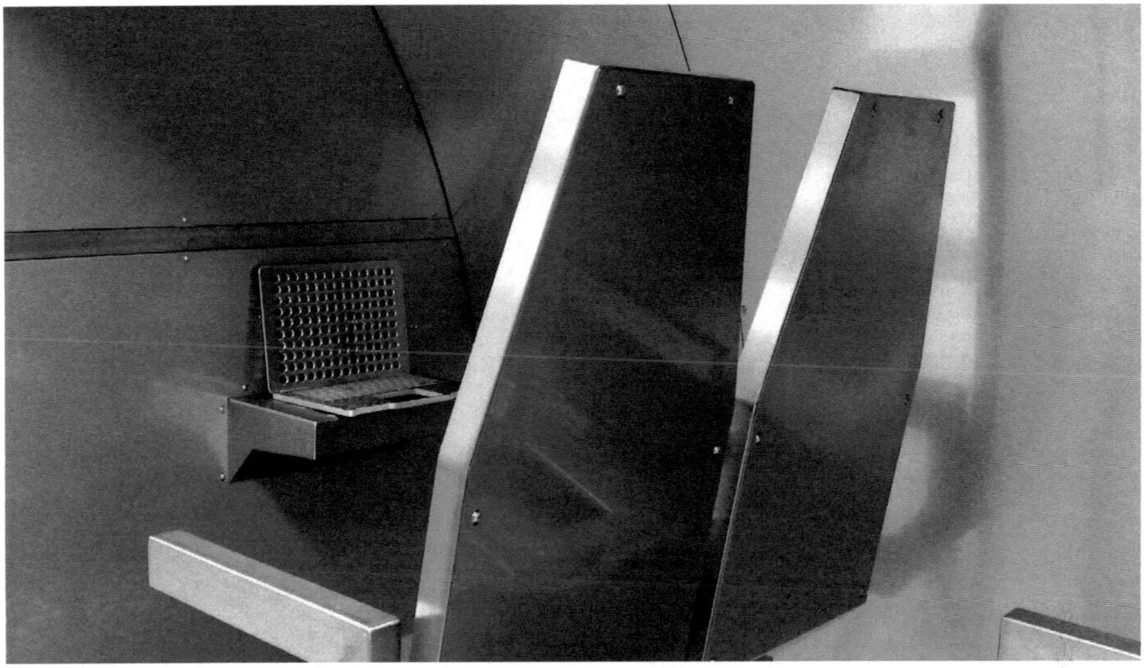

Initial D parameters settings

✓ Lighting:	50% (dimmed)
✓ Smoke density	75% (high)
✓ Cabin panic sound	ON
✓ Burning smell	ON
✓ Difficulty level	Level 4 (hard)
✓ Total training time	1,5 minutes(short)
✓ Relight time	6 seconds(short)

FLAME AVIATION

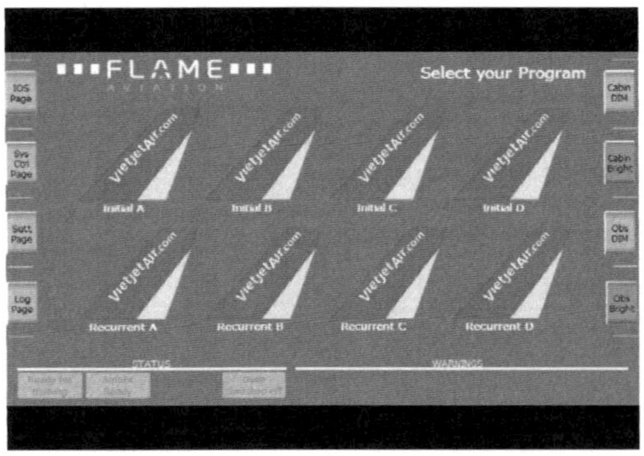

1. Present different scenarios for initial training or for recurrent training
2. Incidents from the line can be incorporated
3. Scenarios run automatically
4. Everyone is presented the same scenario: standardization of training

3. OBJECTIVELY MEASURING PERFORMANCE

23

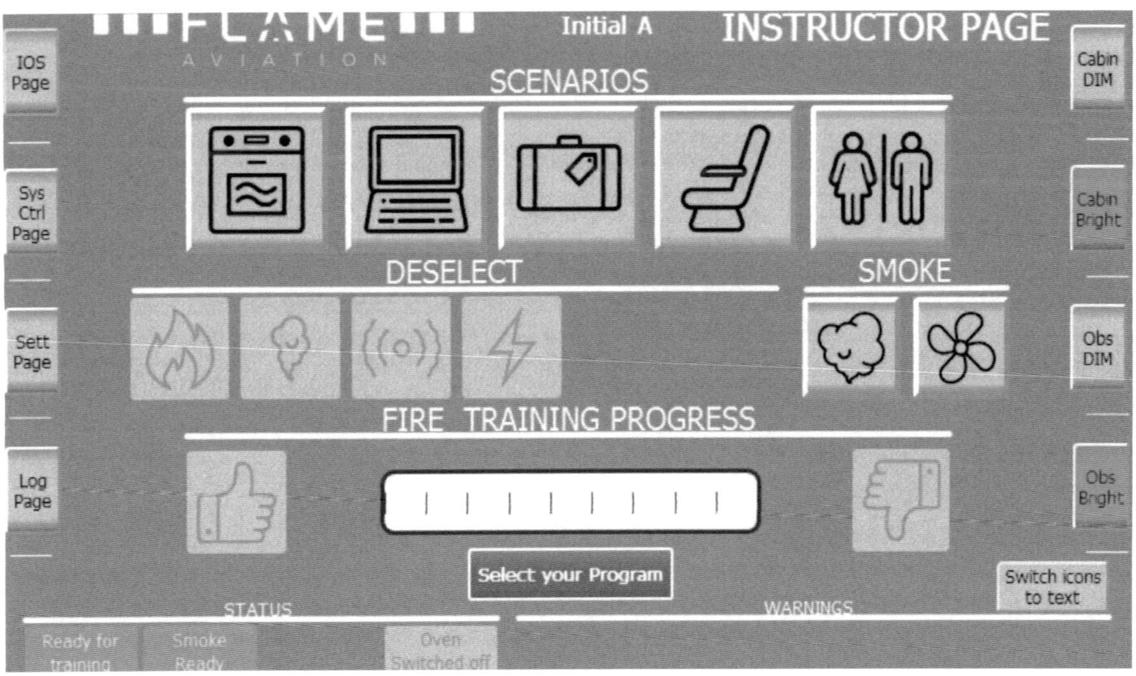

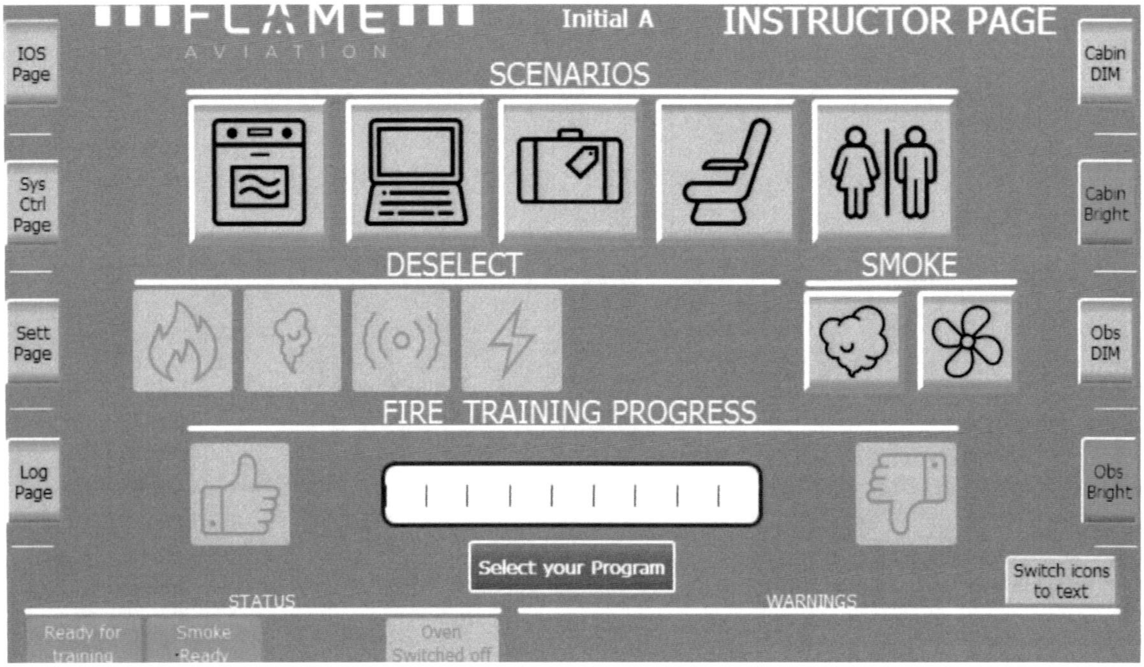

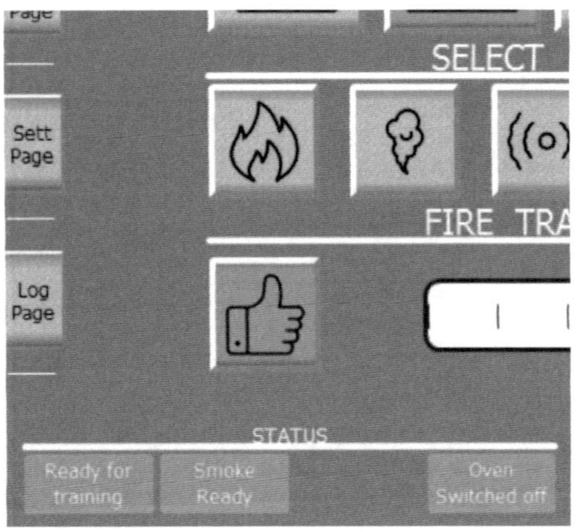

1. Sensors are measuring the performance
2. Difficulty and circumstances are configurable
3. Everyone is assessed by the same criteria: *objectively measuring performance*
4. Possibility to link results to your LMS

FLAME AVIATION 26

Questions

- Is the traditional curriculum still effective?
- Can we complement stand-up instruction with CBT?
- Can we use VR?
- Can we standardize our training?
- Can performance be measured?

FLAME AVIATION

27

Thank you, for your attention!

FLAME AVIATION

28

Human
Trafficking
What do I see?
What do I do?
What do I say?
AVION LLC
TRAINING

Awareness
and training
are different.
Awareness = Recognition
Training = Empowerment

You are in a unique position

- Aviation is one of the primary modes of transportation used by traffickers.
- Cabin crew are front-line personnel (face-to-face).
- Are in a unique position to be able to recognize visible signs of trafficking.
- Have enough time with passengers to be able to observe questionable situations.

Scenarios

Remember, Do No Harm

- Safety is your first concern.
- If you decide to take action, such as reporting it to authorities, do so in a way that will not endanger yourself or others.
- Do not confront suspected traffickers or transporters.
- Do not try to rescue possible victim.
- Do not display concern or alarm.

Thank You

What do you see?
Do you know what to do?
Do you know who to tell?

Find out the hotline numbers for your country and keep them handy when you fly.

AAI Tip Line
Call **802.872.6199**
(tolls apply) in any country in the world, 24 hours

Nat'l Human Trafficking Resource Center
(for victim's services 888.373.7888 – toll free in US
In the US can also call 911

Thank you for your support in stopping
Human Slavery

For more information email me at:
aviontraining1@gmail.com

And check out AAI's Tip Line App
Airlineamb.org/tip-line-app/

AVION LLC TRAINING

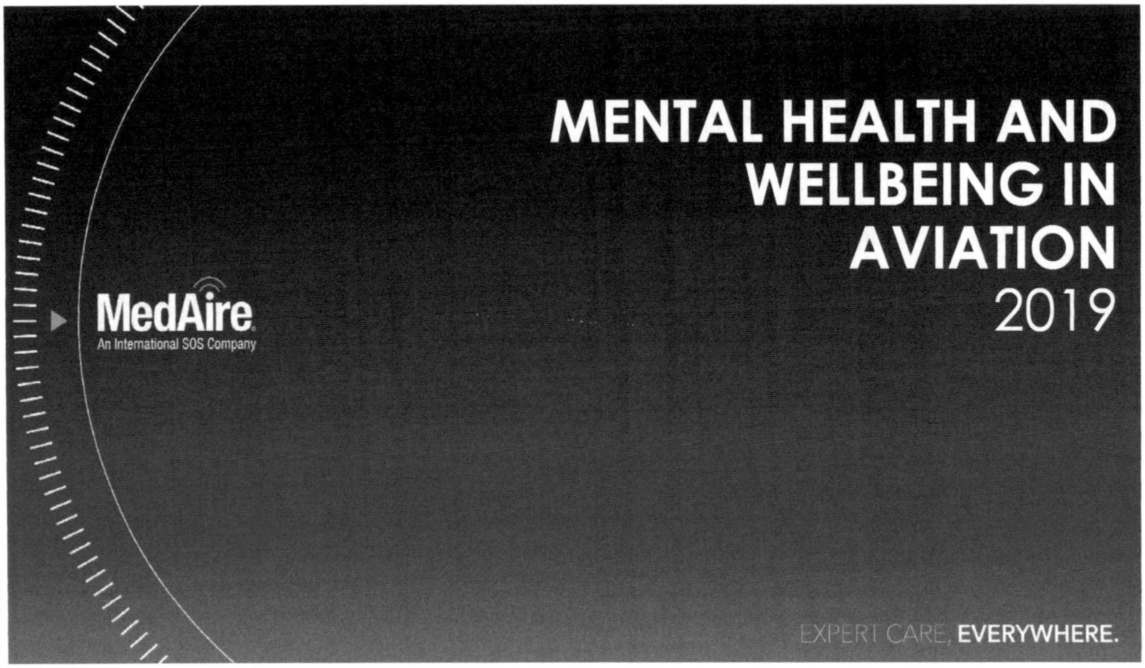
MENTAL HEALTH AND WELLBEING IN AVIATION
2019
MedAire
An International SOS Company
EXPERT CARE, EVERYWHERE.

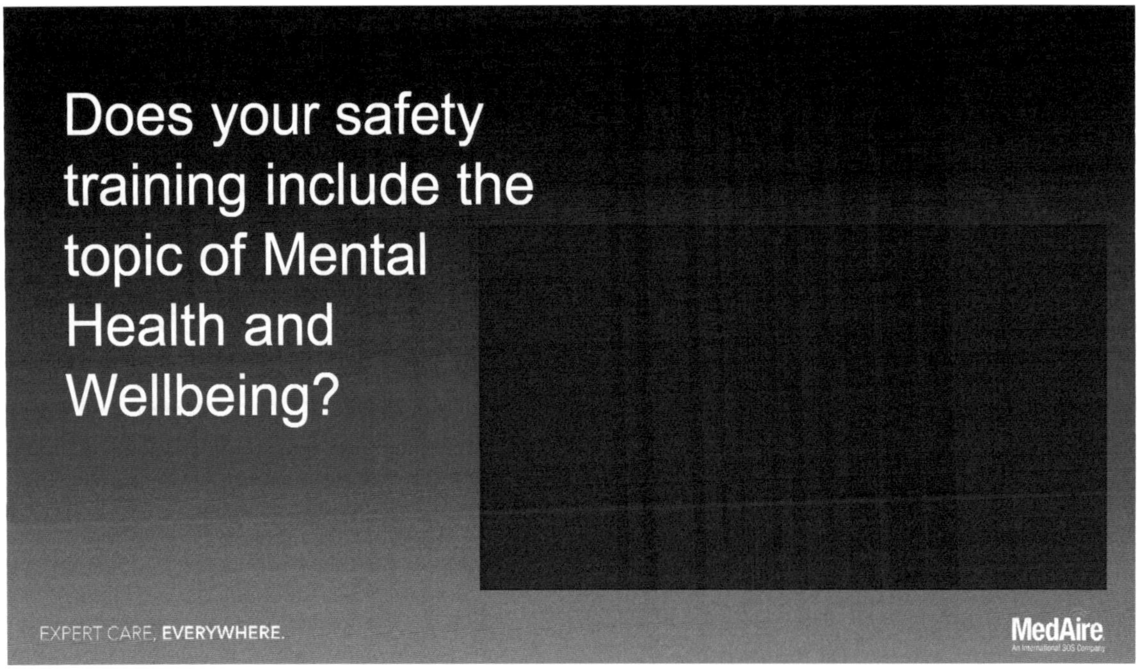
Does your safety training include the topic of Mental Health and Wellbeing?
EXPERT CARE, EVERYWHERE.
MedAire

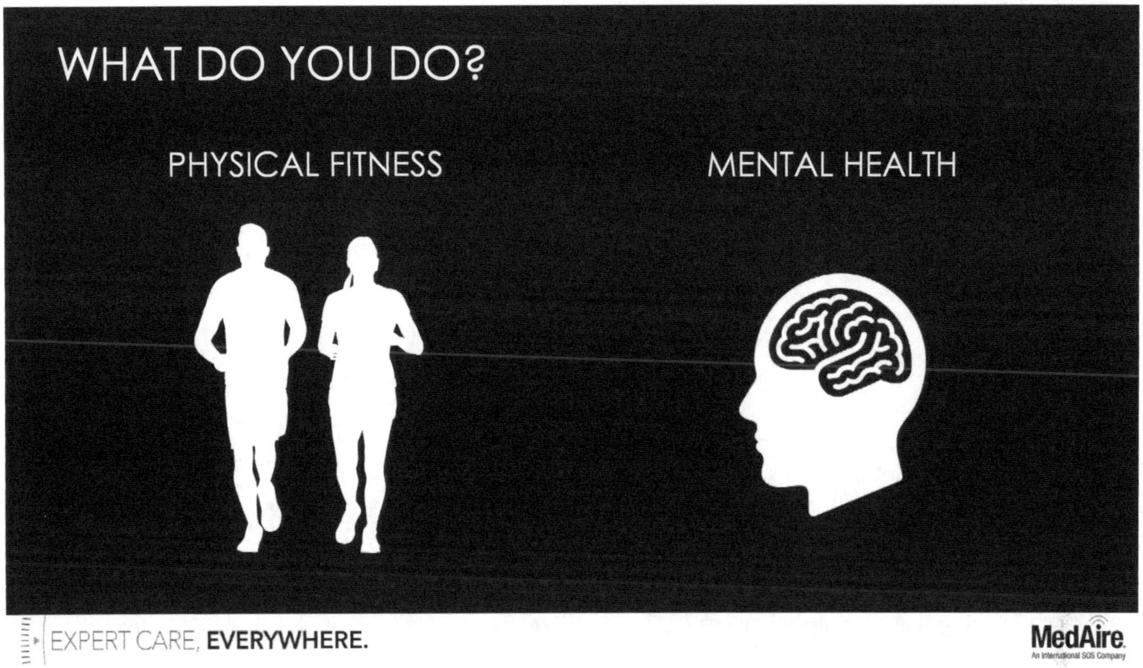
WHAT DO YOU DO?
PHYSICAL FITNESS
MENTAL HEALTH
EXPERT CARE, EVERYWHERE.
MedAire
An International SOS Company

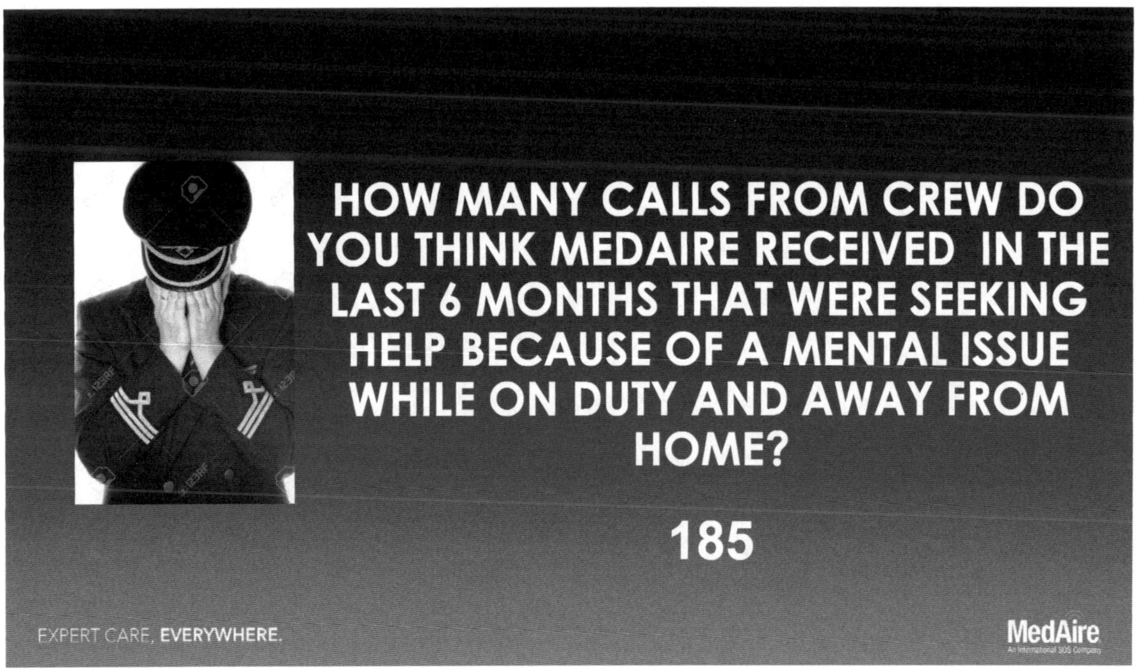
HOW MANY CALLS FROM CREW DO YOU THINK MEDAIRE RECEIVED IN THE LAST 6 MONTHS THAT WERE SEEKING HELP BECAUSE OF A MENTAL ISSUE WHILE ON DUTY AND AWAY FROM HOME?
185
EXPERT CARE, EVERYWHERE.
MedAire
An International SOS Company

SAFETY OF FLIGHT

- What do you think of when you hear "Mental Health In Aviation"?
- Unique environment / workplace - unique position (safety sensitive)

EXPERT CARE, **EVERYWHERE.** MedAire. An International SOS Company

INDUSTRY DRIVERS

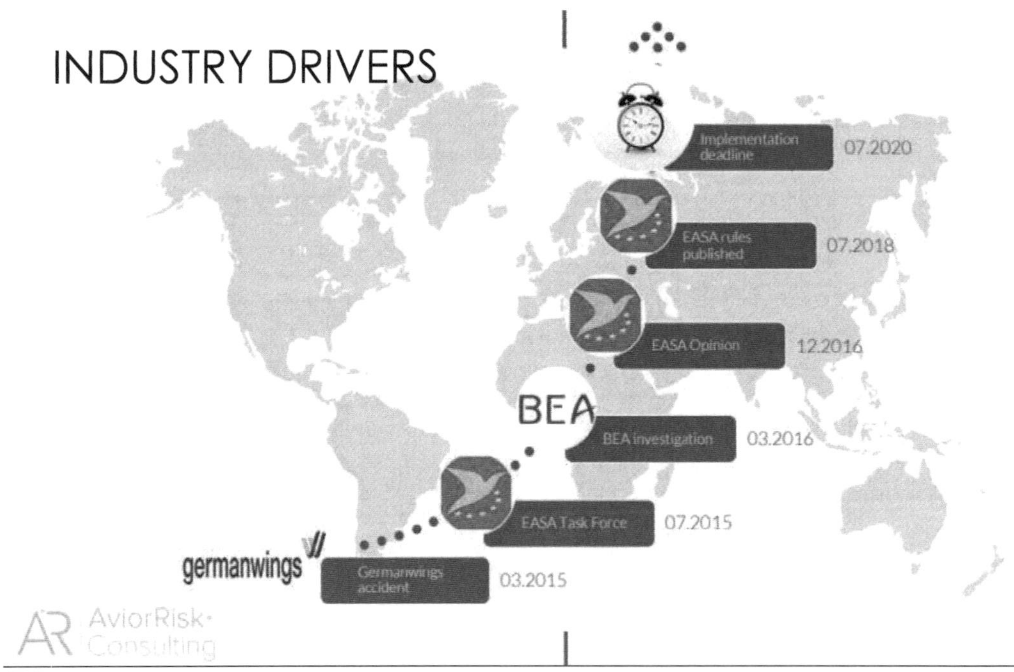

TRIGGERS IMPACTING EMPLOYEE WELLBEING

NORMAL	MOBILITY-RELATED	CRISIS-RELATED
• Stress • Work Overload • Harassment • Bullying • Substance use • Loss of control • Isolation • Depression • Anxiety	• Poor sleep and diet • Lack of support network • Separation from loved ones • Culture clash • Work overload • Monotony/boredom	• Conflict or violence • Colleague death • Accident • Incident at home • Severe weather

SOURCE: International Corporate Health Leadership Council (www.ichlg.org) founded by International SOS

APPROACH TO CREW SUPPORT

Emotional Support

Practical Support

Physical Support

EXPERT CARE, **EVERYWHERE.** MedAire An International SOS Company

MENTAL HEALTH AND WELLBEING CONTINUUM OF CARE

CREW → HOME → HOME → WORKING ON THE ROAD → HOME

PREVENTION	EMPLOYER	24/7 ACCESS	EMPLOYER
▶ TRAINING ▶ ASSESSMENT TOOLS RESOURCES	▶ EAP SERVICES	▶ ASSESSMENT FIT TO OPERATE	▶ EAP SERVICES

Does your program cover Mental Health?
Does your crew have access 24/7 – globally?

EXPERT CARE, **EVERYWHERE.** MedAire

YOUR TRAINING SHOULD INCLUDE

EXPERT CARE, **EVERYWHERE.** MedAire

BEGIN WITH THE BASICS

- Include awareness training – crew should recognize the triggers. (use data)
- Include resources for the crew in the training (your provider numbers)
- Review Safety Event Reports to track trends and adjust training as needed.
- Teach some basic techniques

EXPERT CARE, **EVERYWHERE.** MedAire An International SOS Company

PSYCHOLOGICAL FIRST AID (PFA)

The Eight Core Actions of Psychological First Aid:

1. Contact and engagement
2. Safety and comfort
3. Stabilization
4. Information gathering
5. Practical assistance
6. Connection with social supports
7. Information on coping
8. Linkage with collaborative services

EXPERT CARE, **EVERYWHERE.** INTERNATIONAL SOS An International SOS Company

QUESTION?

THANK YOU!

MedAire
An International SOS Company

AUTHOR INDEX

Halldale Media Inc.
735 Primera Blvd.
Suite 220
Lake Mary, Florida 32746
USA

CURRAN ASSOCIATES INC.
proceedings.com

ISBN 978-1-7138-1889-2
9 781713 818892

10th Malaysian Road Conference & Exhibition 2018

IOP Conference Series: Materials Science and Engineering
Volume 512

Selangor, Malaysia
29 - 31 October 2018

IOP
Institute of Physics